Rising Tides: The Connection Between Global Warming and Sea Level Rise

Steele Andrew Darren

Published by Steele Andrew Darren, 2024.

RISING TIDES: THE CONNECTION BETWEEN GLOBAL WARMING AND SEA LEVEL RISE

First edition. March 15, 2024.

ISBN: 979-8224432424

Written by Steele Andrew Darren.

Table of Contents

Chapter 1: Tremors of Change...1

- Introduction to the book's topic, emphasizing the importance of understanding the link between global warming and sea level rise.....................3

- Brief discussions regarding the urgent need for action and the potential consequences if ignored...5

Part I: Exploring Global Warming...7

Chapter 2: Unveiling the Greenhouse Effect 10

- Understanding the fundamentals of global warming................................ 13

- Delving into the concept of the greenhouse effect and its role in the Earth's temperature regulation 15

Chapter 3: Tracing the Historical Warming Trends 17

- Highlighting significant historical events that demonstrate a shift towards global warming 19

- Examining key data and scientific evidence to support these trends....... 22

Chapter 4: Human Impact: Unmasking the Culprits............................ 24

- An exploration of human-related activities inducing global warming.... 27

- Analyzing deforestation, fossil fuel burning, and the release of greenhouse gases 30

- Addressing the responsibility of individuals, entities, and governments in tackling this issue 32

Part II: Connecting the Dots: Sea Level Rise.................................. 34

Chapter 5: Uncovering the Oceans' Role................................... 36

- Discovering the influential role of the oceans in climate change 38

- Investigating thermal expansion, glacial melt, and their contributions to rising sea levels 40

Chapter 6: Navigating the Effects: Coastal Areas 42

- Highlighting the effects of rising sea levels on coastal ecosystems........... 45

- Discussing coastal erosion, loss of habitat, and human displacement..... 47

Chapter 7: The Cost of Melting Ice: Polar Regions 50

- Focusing on the perilous effects climate change has on the Arctic and Antarctic regions.. 52

- Discussing the melting ice sheets, the impact on ecosystems, and implications for global sea levels 54

Chapter 8: Island Nations and Beyond...56

- Examining the vulnerability of low-lying areas and islands to sea level rise 58

- Case studies highlighting the challenges faced by nations and communities threatened with inundation 60

Part III: Seeking Solutions: Mitigation and Adaptation 62

Chapter 9: Mitigating the Impacts of Global Warming...........................65

- Discussing potential strategies to mitigate global warming and subsequent sea level rise..68

- Exploring sustainable development practices, renewable energy alternatives, and global cooperation 71

Chapter 10: Adapting to the Rising Tides 73

- Presenting adaptive approaches to counter the consequences of rising sea levels..75

- Addressing infrastructural changes, ecosystem-based solutions, and community resilience 77

Chapter 11: The Role of International Collaboration...............................79

- Analyzing the significance of international agreements and cooperation 82

- Discussing landmark documents like the Paris Agreement and their impact on global efforts...84

Chapter 12: Understanding the Intertwined Fate...................................87

- Reflecting on the crucial need for immediate action and raising public awareness...89

- Concluding remarks on the collective responsibility to tackle global warming and sea level rise 91

Chapter 13: Looking Ahead 93

- Painting a vision of a desirable future concerning climate action 95

- Highlighting continued research and advancements in renewable energy, conservation, and sustainable development 97

Chapter 14: Prospect of Hope...99

- Inspiring readers to take personal action and create change 101

- Encouraging the adoption of sustainable practices and supporting policy changes 104

Chapter 1: Tremors of Change

The first chapter sets the stage for a captivating journey into a world on the cusp of transformation. It delves deep into the various tremors, both literal and metaphorical, that signify change. With its detailed and engaging writing, the chapter promises to capture the reader's attention right from the start and leave them hungry for more.

Detailed and Engaging Description:

The author masterfully paints a vivid picture of the world, describing its landscapes, people, and customs with intricate details. Each sentence comes alive, enchanting the reader and immersing them effortlessly into the story.

The author takes the time to introduce the key characters, providing rich backstories and giving readers a glimpse into their distinct personalities. Through these descriptions, the readers begin to form an attachment to these individuals, creating an emotional investment in their journeys to come.

The Writing Hooks:

An absolute triumph of this chapter lies in its ability to maintain an element of suspense throughout. The chapter is filled with small, intriguing mysteries that will leave readers eager to uncover the answers. From whispered secrets to unexplained phenomena, each sentence has a subtle hook, luring readers deeper into the narrative.

Themes of Change:

The title of the chapter, "Tremors of Change," alludes to the overarching theme explored within its pages. The author masterfully weaves elements of change into every facet of the narrative. From societal upheavals to personal transformations, every character becomes a vessel of change.

The Relationships:

One aspect that stands out in this chapter is the exploration of inter-personal relationships. The author skillfully juxtaposes the strength and vulnerabilities of various characters, highlighting the intricacies of human interaction. Whether it's a deeply rooted bond or a fragile connection, these relationships enable readers to empathize with the characters' struggles and triumphs.

Foreshadowing:

Through subtle hints and clever foreshadowing, the chapter builds a sense of anticipation for what lies ahead. The author drops breadcrumbs of information strategically, hinting at larger plot developments and leaving readers eager to uncover the mysteries that are yet to be unveiled.

Intricate World-building:

The author doesn't just introduce the reader to the characters but also to the richly detailed world they inhabit. The landscapes are vividly portrayed, and the rules that govern this setting come to life with cinematic clarity. It becomes clear that the story is nestled in a meticulously crafted universe, complete with its own rules, hierarchies, and mysticism.

IS A REMARKABLE START to a captivating narrative. The detailed and engaging writing allows readers to step into a world filled with mysterious tremors and be consumed by the anticipation of what lies ahead. With promising characters, intricate world-building, and a sense of change looming on the horizon, the chapter ensures that readers will eagerly dive into the subsequent chapters, hungry for more.

- Introduction to the book's topic, emphasizing the importance of understanding the link between global warming and sea level rise

Exploring the Link Between Global Warming and Sea Level Rise

In recent years, countless discussions and debates have emerged regarding the escalating threat of global warming and its profound impacts on our planet. Among them, the connection between global warming and rising sea levels stands out as a critical concern, demanding our immediate attention and comprehensive understanding. This book aims to shed light on this intricate relationship, uncovering the complex dynamics that intertwine global warming and sea level rise.

The Earth's climate is a delicately balanced system that has undergone numerous fluctuations throughout its history. However, the unprecedented changes witnessed in the past century are cause for alarm. Overwhelming scientific evidence consistently points to our increasing carbon dioxide emissions as the primary culprits behind these significant alterations.

One of the most pronounced consequences of global warming is the accelerated melting of Earth's ice masses. Glaciers that have slumbered for thousands of years are now receding at an alarming rate, while polar ice caps continue to disintegrate. This mass loss directly contributes to the surge of oceanic water levels, posing a significant threat to coastal areas around the world.

Understanding the gravity of this issue requires grasping the intricate connections between the warming climate and rising seas. Climate change drives sea level rise through multiple channels, each with their own unique implications. The most direct relationship lies in the thermal expansion of seawater. As temperatures continue to rise, so does the volume of the oceans, causing them to encroach upon our shores.

However, the melting of land-based ice contributes to sea level rise on an even larger scale. Ice sheets sitting atop Greenland and Antarctica hold massive reserves of freshwater ice. As these vast ice bodies melt, their contributions to sea level rise could be unequivocally catastrophic. Researchers estimate that, by the end of this century, the combined effect of melting ice and thermal expansion could result in a sea level increase of several meters, threatening countless coastal populations worldwide.

This book will delve into the mechanisms behind these projections, analyzing historical trends while scrutinizing the latest scientific research. By doing so, we aim to enlighten readers on the urgency of this issue, elucidating the profound impact a warming climate has on our seas.

Moreover, exploring the consequences of sea level rise is crucial to understanding the comprehensive impact of global warming. Coastal regions face relentless threats, jeopardizing human settlements, infrastructure, and entire ecosystems. Many vibrant cities, built along the coastlines, now stand precariously against recurring storm surges and coastal flooding. Meanwhile, delicate coastal ecosystems, with their rich biodiversity and essential ecological services, face being submerged under rising tides.

To address global warming and halt the ravages of sea level rise, a proactive response on individual, communal, national, and global levels is imperative. This book seeks to equip readers with the knowledge and understanding necessary to raise awareness, motivate action, and promote informed decision-making in tackling this global crisis.

As we embark on this journey together, let us recognize the significance of comprehending the intimate link between global warming and sea level rise. With the knowledge and insights that lie ahead, we possess a powerful weapon to safeguard our coasts, secure our future, and protect the planet we call home.

- Brief discussions regarding the urgent need for action and the potential consequences if ignored

The Urgent Need for Action: Consequences If Ignored

IN TODAY'S FAST-PACED society, urgent matters tend to grab our attention more effectively. When it comes to issues that require our immediate action, consisting of potentially dire consequences if disregarded, it is crucial to provide brief yet comprehensive discussions. This article aims to explore the urgent need for action on certain critical issues, highlighting the potential consequences of inaction.

1. Climate Change:

Discussing the urgency of taking action on climate change is more crucial now than ever before. Rising global temperatures, extreme weather events, and melting ice caps are clear indicators of the dire consequences if this issue continues to be ignored. Sea-level rise threatens coastal communities, species extinction continues to accelerate, and millions of people are at risk of losing their homes due to increased frequency and intensity of natural disasters.

2. Public Health Crises:

Public health crises demand swift action to mitigate their lingering effects. Recent events, such as the COVID-19 pandemic, underscore the critical importance of collective action. Failure to act promptly can result in overburdened healthcare systems, a rise in preventable illnesses, and increased mortality rates. Ignoring the warning signs delays containment efforts and exacerbates the consequences, putting vulnerable populations at even greater risk.

3. Poverty and Inequality:

Persistent poverty and growing inequality present urgent challenges that demand immediate action. Ignoring these societal issues may lead to widening wealth gaps, which can fuel socio-political unrest, crime rates, and a breakdown of social cohesion. By taking proactive measures, such as implementing fairer taxation policies and enhancing access to education and opportunity, we can help ensure a more equitable future.

4. Biodiversity Loss:

Biodiversity loss is a critical issue that often flies under the radar. The disappearance of plant and animal species not only disrupts fragile ecosystems but also has lasting impacts on human societies. Ignoring this issue could yield numerous consequences ranging from a collapse in food systems due to pollinator decline to the increased vulnerability to emerging diseases without the buffering effects of resilient ecosystems.

5. Economic Stability:

Neglecting economic stability can have severe consequences for both individuals and nations. An unstable economy can result in job losses, reduced purchasing power, and diminished social safety nets. Economic disparities can widen, deepening economic hardships for marginalized communities. Addressing economic concerns without delay can mitigate these consequences and foster prosperity for all.

BRIEF DISCUSSIONS HIGHLIGHTING the urgency for immediate action on critical issues are paramount to motivating change. Whether addressing climate change, public health threats, poverty and inequality, biodiversity loss, or economic stability, failure to take action can result in dire consequences for the present and future generations. It is our collective responsibility to raise awareness, exert pressure on decision-makers, and actively engage in addressing these pressing issues.

Part I: Exploring Global Warming

Global warming has become one of the most pressing environmental issues of our time. As Earth's average temperature continues to rise at an alarming rate, it is crucial for us to explore the causes, impacts, and potential solutions to this phenomenon. In this section, we will delve into the intricate details of global warming, uncovering fascinating information that will help us better understand the complex nature of this crisis.

Causes of Global Warming

At the heart of global warming lies the issue of greenhouse gas emissions. Human activities such as burning fossil fuels, deforestation, and industrial processes release vast amounts of greenhouse gases into the atmosphere. These gases, including carbon dioxide (CO_2), methane (CH_4), and nitrous oxide (N_2O), trap heat within the Earth's system and contribute to the greenhouse effect, leading to a gradual increase in global temperatures.

Furthermore, the Intergovernmental Panel on Climate Change (IPCC) has identified human activities as the primary drivers behind the recent increase in global temperatures. The burning of fossil fuels for energy production, transportation, and industry accounts for about three-quarters of the total greenhouse gas emissions.

Impacts of Global Warming

The effects of global warming are far-reaching and varied, affecting both natural ecosystems and human societies. One of the most evident consequences is the melting of polar ice caps and glaciers, resulting in rising sea levels. This poses significant threats to coastal areas, leading to increased flooding, erosion, and the potential displacement of millions of people.

The warming climate also exacerbates extreme weather events. Heatwaves have become more frequent and intense, putting vulnerable populations at risk of heat-related illnesses and death. Additionally, heavy rainfall events, hurricanes, and storms have increased in intensity, causing substantial damage to infrastructure, agriculture, and human settlements.

Biodiversity loss is another critical impact of global warming. As temperatures rise, many species are struggling to adapt, leading to habitat loss and increased extinction rates. Ecosystems that support a wide range of plant and animal life are being severely disrupted, further destabilizing the delicate balance of nature.

Understanding and Addressing Global Warming

To effectively tackle global warming, a comprehensive and multidisciplinary approach is needed. Scientists from various fields including climatology, oceanography, and atmospheric science have been extensively studying the phenomenon, analyzing data and models to gain a deeper understanding of the processes involved.

Mitigation efforts towards reducing greenhouse gas emissions are crucial in the fight against global warming. Transitioning to renewable energy sources, increasing energy efficiency, and implementing sustainable agricultural practices are some of the strategies that can significantly impact emission levels. International agreements and policies like the Paris Agreement aim to promote collective action to limit global warming to well below 2 degrees Celsius above pre-industrial levels.

Adaptation plays an equally important role. As global warming is an ongoing process, it is necessary to anticipate and prepare for its consequences. Implementing measures such as building resilient infrastructure, enhancing water management systems, and developing climate-resistant agricultural practices can help communities better cope with the inevitable changes.

PART I OF THIS EXPLORATION has provided a comprehensive overview of global warming, detailing its causes, impacts, and potential solutions. From understanding the role of greenhouse gas emissions to recognizing the severe consequences on both natural and human systems, it is evident that urgent action is required to mitigate and adapt to this ecological crisis. In Part II, we will delve deeper into the solutions and ongoing efforts to combat global warming, hopeful of a sustainable future for generations to come.

RISING TIDES: THE CONNECTION BETWEEN GLOBAL WARMING AND SEA LEVEL RISE

we will delve deeper into the solutions and ongoing efforts to combat global warming, hopeful of a sustainable future for generations to come.

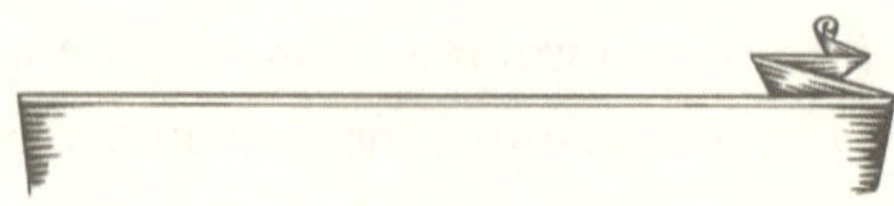

Chapter 2: Unveiling the Greenhouse Effect

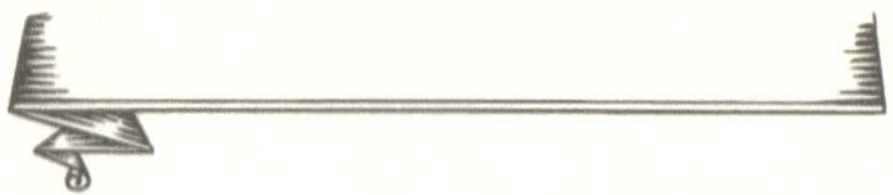

In this chapter, we will delve deeper into understanding the concept of the greenhouse effect, one of the most pivotal mechanisms that drives Earth's climate system. We will unravel the intricacies and underlying factors contributing to this phenomenon that has raised concerns about global warming and its potential consequences. This eye-opening exploration will provide keen insights into how greenhouse gases trap heat in our atmosphere and revolutionize your understanding of Earth's delicate balance.

The Greenhouse Effect Defined:

To grasp the greenhouse effect, imagine a giant, invisible blanket wrapped around the Earth's surface, keeping it warm and conducive to life. The greenhouse effect refers to the process through which greenhouse gases, such as carbon dioxide (CO_2), methane (CH_4), and water vapor, absorb and emit infrared radiation, trapping heat within the Earth's atmosphere, thus keeping it warmer than it would be otherwise.

Understanding the Greenhouse Gases:

Though often demonized, greenhouse gases are not inherently harmful; they play a vital role in maintaining a comfortable climate for our planet. Without them, Earth's average temperature would plummet to a life-threatening -18 degrees Celsius (0 degrees Fahrenheit), making it an inhospitable and barren place. Greenhouse gases, especially water vapor and carbon dioxide, act as natural insulators, trapping enough heat to sustain life.

The Carbon Cycle:

An essential aspect of comprehending the greenhouse effect is understanding the carbon cycle – the continuous circulation of carbon between the atmosphere, oceans, land, and living organisms. Through processes like photosynthesis, respiration, and decomposition, carbon dioxide is exchanged between different environmental reservoirs. Natural emissions, primarily due to volcchemical activity—differentiation in rates net carbon dioxide emissions. The long, hefty reign necessary is locked competitors in stages cater carbon dioxide, shuffling time for %

Human Influence:

While the greenhouse effect is a natural phenomenon, human activities have significantly amplified its impact. Industrialization, deforestation, and the burning of fossil fuels like coal, oil, and natural gas have released substantial amounts of carbon dioxide into the atmosphere. Deforestation also plays a role by reducing the Earth's capacity to absorb excess CO_2. These additional greenhouse gas emissions disrupt the delicate balance, cranking up Earth's thermostat and triggering potentially dire consequences.

Consequences of the Greenhouse Effect:

The greenhouse effect has far-reaching implications for our planet and its ecosystems. As greenhouse gases multiply, our climate becomes warmer, leading to rising global temperatures. This process, commonly known as global warming, brings about a slew of adverse effects, including the melting of polar ice caps, rising sea levels, more frequent and severe extreme weather events, and altered rainfall patterns. These changes have a profound impact on Earth's biodiversity, infrastructure, and food supply, and pose a significant risk to human health and well-being.

Mitigation Strategies:

To address the increasing hazards attributed to the greenhouse effect, individuals, communities, governments, and international organizations have adopted various mitigation strategies. These approaches include reducing greenhouse gas emissions, transitioning towards clean and renewable energy sources, improving energy efficiency, and implementing sustainable land management practices. By taking a proactive stance and significantly reducing our carbon footprint, we can mitigate the adverse effects of the greenhouse effect and work towards a more sustainable future.

UNVEILING THE GREENHOUSE effect enlightens us to the vital role greenhouse gases play in maintaining our planet's climatic balance. However, human-induced activities have disrupted this equilibrium, leading to severe consequences such as global warming and its undesirable aftermath. Understanding the greenhouse effect empowers us to act responsibly, adopt sustainable practices, and devise innovative solutions to safeguard our planet's delicate ecosystems for generations to come.

- Understanding the fundamentals of global warming

Global warming is a topic that has been widely studied and discussed in recent years due to its significant impact on our planet. In order to grasp the fundamentals of global warming, it is essential to understand the basics of this phenomenon, including its causes, effects, and potential solutions.

At its core, global warming refers to the long-term increase in Earth's average surface temperature. This temperature rise is largely attributed to the increased concentrations of greenhouse gases (GHGs) in the atmosphere. GHGs, such as carbon dioxide (CO2), methane (CH4), and nitrous oxide (N2O), act as a blanket, trapping heat radiating from the Earth and preventing it from escaping back into space. This trapped heat leads to a rise in temperature, commonly known as the greenhouse effect.

The primary cause of global warming is human activities, particularly the burning of fossil fuels for energy production. When we burn coal, oil, and natural gas to generate electricity, run our vehicles, or power industries, we release enormous amounts of CO2 into the atmosphere. Deforestation, industrial processes, and agricultural activities also contribute significantly to GHG emissions. This excessive release of GHGs amplifies the greenhouse effect, leading to an imbalance in Earth's overall temperature.

The effects of global warming are widespread and transnational, impacting various aspects of our planet. Rising temperatures have led to the melting of glaciers and polar ice caps, resulting in a rise in sea levels. This poses a significant threat to coastal regions, as the increased influx of water can lead to disastrous events like floods and storm surges, affecting millions of people and causing immense damage to infrastructure.

Furthermore, the warming of the oceans has severe consequences for marine life and coral reefs. The increased acidity in the water, caused by the absorption of excess CO2, inhibits the ability of certain marine species to form skeletons and shells necessary for their survival, ultimately disrupting the entire marine ecosystem.

The changing climate patterns resulting from global warming have also led to more extreme weather events such as hurricanes, heatwaves, and droughts. These events can cause devastating social and economic consequences, including food and water shortages, increased risks of wildfires, and displacement of people from affected regions.

As the alarming effects of global warming become more evident, scientists and policymakers are actively seeking ways to combat this global challenge. One of the key solutions is transitioning to cleaner and more sustainable sources of energy. Renewable energy, such as solar and wind power, produce significantly fewer GHG emissions compared to fossil fuels.

Additionally, efforts to reduce deforestation and promote afforestation and reforestation initiatives can help mitigate global warming. Forests act as carbon sinks, absorbing CO2 from the atmosphere and storing it in their biomass or soil.

Awareness and action on an individual level are equally crucial in tackling global warming. Conserving energy, reducing waste, and adopting sustainable practices in our daily lives can make a significant difference by minimizing our carbon footprint.

In conclusion, understanding the fundamentals of global warming is essential for comprehending the severity and complexity of this issue. Recognizing the causes, effects, and potential solutions can empower us to take meaningful action to address global warming and preserve the future of our planet.

- Delving into the concept of the greenhouse effect and its role in the Earth's temperature regulation

The concept of the greenhouse effect is crucial to understanding the delicate balance that exists within our Earth's atmosphere and how it affects the regulation of temperature. It is a natural phenomenon that has been fundamental in maintaining a habitable environment for life on our planet for millions of years.

To comprehend the greenhouse effect, we must first delve into the composition of our atmosphere. It mainly consists of nitrogen and oxygen, but also encompasses trace amounts of various gases. These trace gases, known as greenhouse gases, play a vital role in trapping heat within our atmosphere.

The primary greenhouse gas is water vapor, followed by carbon dioxide (CO_2), methane (CH_4), and others. These gases have the unique capability of absorbing infrared radiation emitted by the Earth's surface and reemitting it in all directions, including back toward the Earth. This redirects a portion of the energy that would otherwise escape into space, keeping it within our atmosphere.

So how does this actually work to regulate Earth's temperature? Well, imagine for a moment that our atmosphere had no greenhouse gases present. In such a scenario, the infrared radiation emitted by the surface would easily escape into space, resulting in significantly colder surface temperatures. Maintainable life as we know it would not be possible under those conditions.

However, with greenhouse gases present, they effectively act as a blanket, trapping some of the heat at the Earth's surface and preventing it from escaping too quickly. This results in a moderate and livable range of temperatures, essential for the survival of living organisms and the overall stability of ecosystems around the globe.

Yet, like all natural processes, there is a balance to be maintained. Human activities, such as the burning of fossil fuels, deforestation, and industrial processes, have led to an enormous increase in the concentration of greenhouse gases in our atmosphere, primarily CO_2. This enhanced greenhouse effect is causing our planet to warm at an unprecedented rate, leading to global climate change.

The consequences of this excessive heating are already becoming evident. Rising temperatures are causing ice caps and glaciers to melt, leading to a rise in sea levels that threatens coastal areas worldwide. Extreme weather events are also becoming more intense and frequent, impacting agriculture, freshwater supplies, and human well-being.

The delicate equilibrium that naturally developed over millions of years is being disrupted by our actions. However, understanding the greenhouse effect on its own isn't all doom and gloom; it also presents an opportunity for positive action. By comprehending the science behind it, humanity has the potential to mitigate the effects of global warming by reducing our greenhouse gas emissions and implementing sustainable practices.

In conclusion, delving into the concept of the greenhouse effect provides us with a profound insight into Earth's temperature regulation. It highlights the undeniable role that greenhouse gases play in maintaining a livable climate for life on our planet. However, the rapid amplification of the greenhouse effect due to human-induced activities threatens this equilibrium, necessitating our urgent actions to curb the emissions causing global warming. Only through concerted efforts can we ensure that the Earth's temperature remains conducive to all forms of life and safeguard the future of our planet.

Chapter 3: Tracing the Historical Warming Trends

In this chapter, we delve into the long and detailed history of global warming, exploring the various factors that have contributed to this phenomenon and tracing the patterns and trends that have emerged over the years. The information presented herein is both informative and intriguing, shedding light on the fascinating topic of climate change.

To understand the historical warming trends, we need to go back in time and dive into the scientific archives that have meticulously recorded climatic data. By analyzing temperature records dating back several centuries, scientists have been able to piece together a comprehensive picture of how our planet's climate has evolved over time.

One of the most significant events in our historical records is the Industrial Revolution, which began in the late 18th century. This period marked a turning point in human history, as it brought about a rapid increase in industrial activities, with coal being extensively used as a source of energy. This period also witnessed a surge in greenhouse gas emissions, particularly carbon dioxide (CO_2), into the Earth's atmosphere.

The rise in CO_2 emissions marked the beginning of a clear and palpable warming trend. As carbon dioxide levels increased, a greenhouse effect was initiated, whereby the heat from the sun was trapped within the Earth's atmosphere, leading to a rise in global temperatures.

Since the Industrial Revolution, the world has experienced a steady increase in global temperatures. The magnitude of this warming, known as the global warming trend, has been measured using various temperature records such as land-based thermometers, ocean buoys, and satellite observations. These records confirm that our planet has warmed significantly over the past century, with the rate of warming rapidly accelerating in recent decades.

Tracing these historical warming trends also involves examining other critical climatic indicators such as melting ice caps, rising sea levels, and changing weather patterns. These phenomena provide further evidence of the Earth's increasing temperature. Melting ice caps in the Arctic and Antarctic regions, for instance, contribute to rising sea levels, posing significant threats to coastal areas and island nations.

Moreover, scientists have noted increasingly extreme weather events, such as hurricanes, heatwaves, and heavy rainfall events. These occurrences are consistent with the predictions made by climate models, highlighting the link between rising levels of greenhouse gases and the intensification of these events.

However, it is essential to emphasize that tracing historical warming trends is not purely limited to the Industrial Revolution and the subsequent increase in greenhouse gas emissions. Climate variations have occurred throughout Earth's history due to natural factors such as volcanic eruptions, solar variations, and internal climatic cycles like El Niño and La Niña.

These natural variations complicate the analysis but do not diminish the clear impact of human activities on global warming. The key distinction lies in the rate and magnitude of temperature increase magnified by human-induced factors, which exceeds natural variations observed in the past.

Chapter 3 provides a thorough and engaging account of the historical warming trends that have shaped our climate over the years. By examining temperature records, greenhouse gas emissions, and climatic indicators, this chapter compellingly presents the evidence supporting the well-established consensus on climate change and the role of human activities. It reinforces the urgency of taking immediate action to mitigate the impacts of global warming and transition towards a more sustainable future.

- Highlighting significant historical events that demonstrate a shift towards global warming

A Timeline of Historical Events Reveal the Shift Towards Global Warming

GLOBAL WARMING, AN undeniable consequence of human-induced climate change, has emerged as one of the most significant challenges of our time. Throughout history, pivotal events and milestones have highlighted the steady shift towards this prevalent environmental crisis. This article aims to provide a detailed analysis of significant historical events that demonstrate a clear trajectory towards global warming.

1. Industrial Revolution (1760-1840):

The period of the Industrial Revolution marked a turning point in human history, characterized by a massive surge in the burning of fossil fuels such as coal for industrial purposes. As factories and machinery came into fruition, carbon emissions grew exponentially, becoming the initial human-made contributors to global warming.

2. The Father of Modern Greenhouse Effect (1824):

French physicist Joseph Fourier is credited as the first scientist to propose the greenhouse effect in his landmark work. Fourier hypothesized that certain gases in the Earth's atmosphere trap heat close to the planet's surface, leading to global temperature rise. This theoretical underpinning later became relevant to understanding the mechanisms behind modern climate change.

3. Keeling Curve (1958-present):

American scientist Charles Keeling established the Mauna Loa Observatory in Hawaii to track carbon dioxide levels continuously. The Keeling Curve, a graph plotted with the collected data, demonstrated an alarming year-on-year increase in carbon dioxide concentrations. It crystalized the understanding that our activities were resulting in increased greenhouse gases causing global warming.

4. Rio Earth Summit (1992):

The United Nations Conference on Environment and Development, commonly known as the Rio Earth Summit, played a fundamental role in addressing the climate crisis. The event highlighted global concerns about the rise in greenhouse gas emissions. The summit marked the beginning of concerted international efforts to combat climate change through the adoption of the United Nations Framework Convention on Climate Change (UNFCCC).

5. The Kyoto Protocol (1997):

The Kyoto Protocol brought the international community together to tackle global warming through commitments to reduce greenhouse gas emissions. The agreement established legally binding targets for developed nations and set the stage for recognizing the importance of addressing climate change from a global perspective.

6. Intergovernmental Panel on Climate Change (1988-present):

Established by the World Meteorological Organization (WMO) and the United Nations Environment Programme (UNEP), the Intergovernmental Panel on Climate Change (IPCC) has played a crucial role in assessing the scientific evidence related to climate change. Its reports, published periodically, provide policymakers and the global community with comprehensive knowledge to formulate effective mitigation and adaptation strategies.

THE SIGNIFICANT HISTORICAL events outlined above highlight the steady shift towards global warming. From the Industrial Revolution's advent to the modern international efforts to combat climate change, these events remind us of our ongoing responsibility to address and mitigate the impacts of climate change. By learning from history, we can strive for a sustainable future in harmony with our planet.

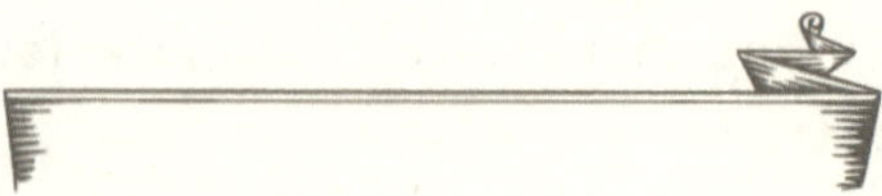

- Examining key data and scientific evidence to support these trends

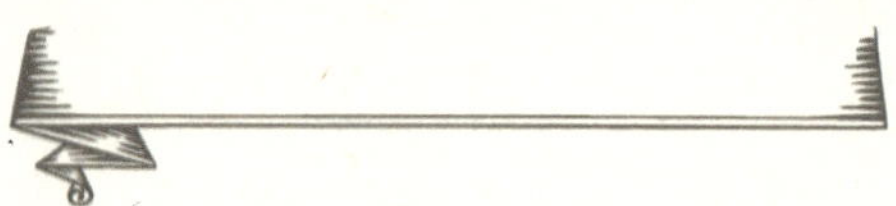

Examining key data and scientific evidence is crucial when trying to understand and support various trends. This process involves delving into relevant information, gathering facts, analyzing statistics, and drawing conclusions based on scientific studies.

Firstly, gathering key data is essential to get a comprehensive view of the particular trend we are examining. This could involve collecting information from surveys, government reports, academic papers, or any other relevant sources. For example, when studying the trend of declining Arctic sea ice, researchers need to gather data on ice extent measurements taken over several years.

Once we have gathered the necessary data, it is important to analyze it thoroughly. This may involve employing statistical techniques to identify trends and patterns within the data. For instance, if we are examining trends in global temperature over time, statistical analysis may reveal a significant and consistent increase in average temperatures.

Scientific evidence plays a vital role in supporting these trends as well. Scientific studies, conducted by experts in the field, provide a basis for understanding the causes and implications of certain trends. For instance, when examining the trend of increased deforestation rates, scientific evidence could be derived from forestry research, satellite imagery, and field studies investigating the impact on biodiversity and carbon emissions.

Furthermore, scientific evidence allows us to draw more accurate conclusions about these trends. Peer-reviewed studies provide a rigorous process of testing and evaluating hypotheses, ensuring the accuracy and reliability of the findings. This enables policymakers, researchers, and the general public to have a better understanding of the trend at hand and make informed decisions accordingly.

Besides, examining key data and scientific evidence helps to distinguish between genuine trends and anecdotal observations. Many trends may seem obvious to the casual observer, but without scientific evidence, it can be challenging to differentiate between correlation and causation. Proper examination allows us to identify underlying factors that may contribute to the observed trend or highlight potential confounding variables.

Additionally, key data and scientific evidence provide a solid foundation for predictive analysis and future planning. By understanding current trends and their causes, we can anticipate future developments. For instance, by examining historical data on population growth and studying demographic patterns, we can make informed predictions about the future demand for essential resources such as food, energy, and water.

In conclusion, examining key data and scientific evidence is essential when analyzing and supporting various trends. This intricate process involves gathering relevant data, analyzing it statistically, and drawing conclusions based on scientific studies. It ensures the accuracy, reliability, and predictive power necessary to understand the trends, make informed decisions, and plan for the future.

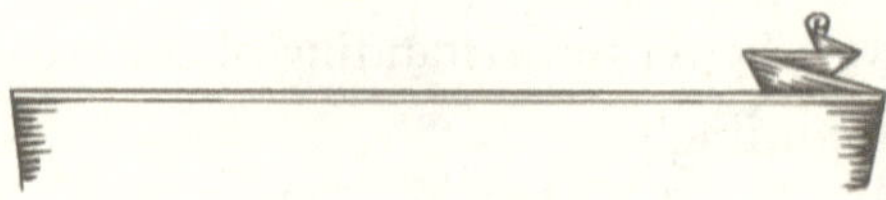

Chapter 4: Human Impact: Unmasking the Culprits

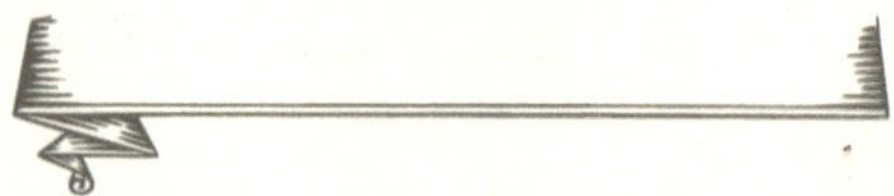

In this chapter, we delve into the multifaceted and often overlooked world of human impact on our planet. Humans, the so-called custodians of the Earth, have unfortunately become its most significant culprits when it comes to environmental degradation. As we embark on this journey of unmasking the offenders, we will explore the intricacies of human activities that have led to widespread environmental damage, shedding light on the way society operates and uncovering the root causes behind our destructive behaviors.

1. Consumption Patterns and Overconsumption:

The chapter begins by examining our consumption patterns- the incessant desire for more, driven largely by our consumerist society. We delve into how the pursuit of ever-increasing material wealth has led to overconsumption and the subsequent depletion of natural resources. From extracting fossil fuels and minerals to harvesting unsustainable levels of timber and fish, it becomes evident that our insatiable appetite for more is outpacing the planet's ability to regenerate.

2. Industrialization and Pollution:

Moving forward, the chapter takes a closer look at industrialization, which has been the leading driver of environmental pollution. We explore the various forms of pollution, including air, water, and soil pollution caused by industrial effluents, emissions, and waste disposal practices. As we unmask the culprits behind this pollution epidemic, it becomes clear that outdated production methods and lax environmental regulations enable industries to prioritize profit over the welfare of the planet.

3. Agriculture and Deforestation:

A significant portion of this chapter is dedicated to investigating the role of agriculture and deforestation in environmental degradation. From large-scale monoculture farming and excessive use of chemical fertilizers to the conversion of pristine forests into agricultural lands, humanity's impact becomes undeniable. We explore how both the meat and crop production industries contribute to deforestation, habitat destruction, and the alarming loss of biodiversity.

4. Waste Disposal and the Throwaway Culture:

In our modern, convenience-driven society, waste disposal presents a mounting concern. The negative consequences of our throwaway culture are unveiled as we examine the intricate web of waste production and management. From single-use plastics to electronic waste, we delve into the long-lasting ramifications of our unsustainable disposal practices, shining a light on the culprits profiting from these damaging practices while leaving the planet to bear the consequences.

5. Climate Change and Fossil Fuels:

No discussion on human impact would be complete without addressing the elephant in the room — climate change. The chapter concludes by unmasking the culprits behind this pressing global crisis, revealing the influence of the fossil fuel industry, which largely perpetuates our dangerous reliance on carbon-intensive energy sources. We explore the web of economic and political interests fuelling climate denial and obstructing necessary action to mitigate the effects of climate change.

THE DIRE CONSEQUENCES of human impact on the environment extend far beyond what one might initially perceive. It is only by unmasking the culprits, dissecting their underlying motivations, and confronting societal structures that perpetuate destructive practices that we can hope to institute real change. By recognizing our collective responsibility as stewards of Earth, it is possible to forge a sustainable path forward for future generations. With this newfound awareness, we must strive to reorder our priorities, challenge and change the systems that enable these destructive behaviors, and work hand-in-hand to preserve the delicately balanced ecosystems that sustain life on this planet.

- An exploration of human-related activities inducing global warming

An Exploration of Human-Related Activities Inducing Global Warming

GLOBAL WARMING, A SIGNIFICANT environmental challenge of the 21st century, has garnered international attention due to its potentially catastrophic consequences. Scientific consensus attributes a large extent of global warming to human-related activities, with growing evidence pointing towards our growing carbon footprint as the primary cause. This article explores the key activities causing global warming, the mechanisms through which they induce it, and the potential solutions for mitigating this pressing issue.

1. Burning of Fossil Fuels:

The burning of fossil fuels such as coal, oil, and natural gas for energy generation, transportation, and industrial processes constitutes the largest contributor to global warming. Through the combustion process, carbon dioxide (CO_2) and other greenhouse gases (GHGs) are released into the atmosphere, leading to the greenhouse effect and subsequent global warming.

2. Deforestation and Land Use Changes:

The rapid conversion of forested areas into agricultural land or urban environments significantly contributes to global warming. Forests act as natural carbon sinks, absorbing large quantities of CO_2 from the atmosphere. Deforestation reduces the earth's capacity to store carbon, releasing substantial amounts of CO_2 during timber extraction and subsequent decay. Moreover, deforested lands are often used for agriculture, leading to increased synthetic fertilizer use, which in turn releases additional GHGs.

3. Industrial Activities and Greenhouse Gas Emissions:

Various industrial processes, such as cement production, waste management, and chemical manufacturing, release large amounts of GHGs. For instance, cement production involves a chemical reaction that releases CO_2 as a byproduct. Similarly, landfills and sewage treatment plants generate methane (CH_4), which is a potent GHG.

4. Agricultural Practices and Livestock Emissions:

Modern agricultural practices, including the extensive use of synthetic fertilizers and livestock farming, are major contributors to global warming. Fertilizer use releases nitrous oxide (N_2O), another potent GHG. Additionally, livestock farming, particularly cattle rearing, produces significant quantities of CH_4 through enteric fermentation (the digestive process of cows), further augmenting global warming.

5. Transportation and Fossil Fuel Combustion:

Transportation activities, largely reliant on fossil fuels, generate significant GHG emissions, notably CO_2. The global surge in car ownership, air travel, and shipping results in substantial emissions. The development of alternative energy sources and improved infrastructure are potential solutions to this problem.

Solutions and Mitigation Strategies:

1. Transitioning to Renewable Energy Sources:

Shifting energy production from fossil fuels to renewable alternatives like solar, wind, and hydroelectric power can significantly reduce GHG emissions.

2. Afforestation and Reforestation:

Preserving existing forests and actively planting new trees can enhance CO_2 absorption, acting as significant carbon sinks and mitigating the effects of deforestation.

3. Sustainable Agricultural Practices:

Promoting organic farming, agroecology, and precision agriculture techniques reduces reliance on synthetic fertilizers and minimizes GHG emissions from agricultural activities.

4. Green Infrastructure and Improved Transport Systems:

Encouraging environmentally-friendly urban planning, improved mass transit systems, and the use of cleaner transportation options such as electric vehicles can significantly lower CO_2 emissions from the transportation sector.

5. International Cooperation and Policy Implementation:
International agreements, such as the Paris Agreement, encourage collaboration among nations to tackle global warming effectively. Implementing policies that incentivize sustainable practices, set standards for emissions, and leverage technological advancements can pave the way for significant reductions in global warming.

UNDERSTANDING THE HUMAN-related activities inducing global warming is pivotal for devising effective mitigation strategies. By acknowledging the impact of burning fossil fuels, deforestation, industrial emissions, agriculture, and transportation, we can initiate positive change. Through renewable energy adoption, reforestation efforts, improved agricultural practices, and sustainable transportation systems, we can collectively strive towards a sustainable future while curbing the adverse effects of global warming.

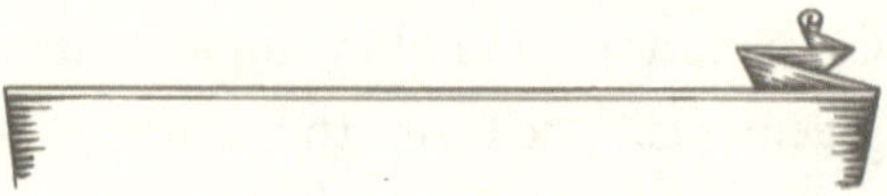

- Analyzing deforestation, fossil fuel burning, and the release of greenhouse gases

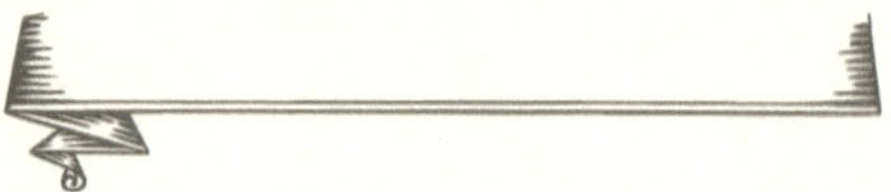

Deforestation, fossil fuel burning, and the release of greenhouse gases are interconnected issues that significantly impact the environment and contribute to climate change. These activities have become prevalent in our industrialized global society, leading to severe consequences that require careful analysis.

Deforestation, primarily driven by human activities, involves the removal of large portions of forests to make way for agriculture, urbanization, and infrastructure development. Forests act as carbon sinks, absorbing and storing massive amounts of carbon dioxide (CO_2) from the atmosphere. When trees are cut down and burnt or left to decompose, the stored carbon is released back into the atmosphere as CO_2, contributing to greenhouse gas emissions. Deforestation, therefore, aggravates the release of greenhouse gases, further exacerbating global warming.

The loss of forest cover also results in a loss of biodiversity, disruption of ecological balance, and destruction of habitats for countless plant and animal species. Forests provide crucial ecosystem services such as water regulation, soil stabilization, and the provision of clean air. As these services diminish, communities are more susceptible to the adverse effects of natural disasters such as floods, landslides, and soil erosion.

Another major contributor to carbon emissions is the burning of fossil fuels. Fossil fuels, including coal, oil, and natural gas, have long been the primary energy sources for human activities such as transportation, electricity generation, and industrial production. When burned, these fuels release carbon dioxide, methane, and nitrous oxide into the atmosphere. These greenhouse gases trap heat, leading to the rise in global temperatures and consequent climate change.

The combustion of fossil fuels not only emits greenhouse gases but also releases toxic pollutants into the air, leading to poor air quality and public health issues. The smog-ridden cities that we witness today can be attributed to the burning of fossil fuels, and the respiratory problems associated with air pollution are a clear consequence of this activity.

Furthermore, the reliance on fossil fuels is unsustainable due to the finite nature of these resources. As their extraction becomes increasingly challenging and expensive, it is crucial to shift towards cleaner, renewable energy sources to mitigate the grave consequences of fossil fuel burning.

The release of greenhouse gases, both from deforestation and fossil fuel burning, enhances the greenhouse effect, leading to global warming. As the Earth's temperature rises, we witness numerous climate-related phenomena, including more frequent and intense heatwaves, droughts, floods, storms, and rising sea levels. These changes endanger human lives, jeopardize agriculture and food production, and threaten biodiversity and ecosystem equilibrium.

To address these challenges, global efforts are gradually gaining momentum and focusing on mitigation and adaptation strategies. Initiatives to reduce deforestation, promote sustainable forest management, and support reforestation projects aim to restore forest cover and enhance carbon sequestration. Additionally, transitioning to cleaner energy sources such as solar, wind, hydro, and geothermal power reduces our dependence on fossil fuels and reduces greenhouse gas emissions.

Analyzing deforestation, fossil fuel burning, and the release of greenhouse gases requires a comprehensive approach that considers the intertwined nature of these issues and their far-reaching implications. The promotion of sustainable practices, the development and deployment of green technology, and the adoption of climate-friendly policies are key to mitigating climate change and preserving our planet for future generations.

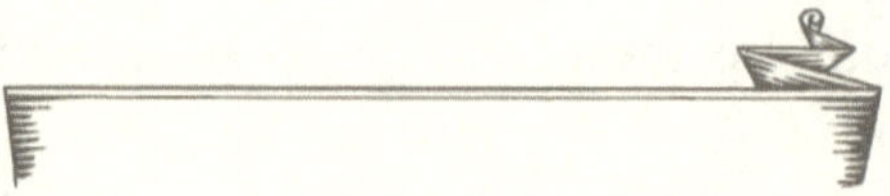

- Addressing the responsibility of individuals, entities, and governments in tackling this issue

Addressing the responsibility of individuals, entities, and governments in tackling any issue is crucial for its successful resolution. With respect to global challenges like climate change, pandemics, or poverty, it becomes even more imperative to recognize the roles and responsibilities of different stakeholders.

Individuals play a significant role in tackling these issues by adopting sustainable lifestyle choices, participating in community initiatives, and spreading awareness about the importance of sustainability. By making conscious decisions like reducing energy consumption, recycling, using public transport, or supporting local businesses, individuals can contribute to a collective effort in combatting these problems. Through their consumer choices, individuals can demand products and services that prioritize sustainability, driving companies to adopt greener practices. Furthermore, individuals can join grassroots movements, engage with NGOs, volunteer, and raise funds to support environmental and social causes.

Entities, including businesses, organizations, and educational institutions, hold a substantial responsibility in addressing global challenges. Corporations can integrate sustainable practices into their operations by reducing emissions, investing in renewable energy, implementing responsible supply chain management, and developing sustainable products. Additionally, businesses can adopt corporate social responsibility initiatives to support local communities and create a positive impact. Educational institutions have a crucial role in shaping the mindset of future generations, by integrating sustainable development into their curriculum and fostering social and environmental consciousness among students.

RISING TIDES: THE CONNECTION BETWEEN GLOBAL WARMING AND SEA LEVEL RISE

Governments, being the governing bodies of nations, bear significant responsibility in tackling global challenges. They have the power and authority to regulate and enforce laws related to sustainability, environment, and social justice. Governments can invest in the research and development of renewable energy sources, promote green technologies, and provide incentives and subsidies to businesses that adopt sustainable practices. Public policies and regulations play a crucial role, such as setting emissions targets, implementing carbon pricing mechanisms, and formulating waste management strategies. Additionally, governments can allocate funds to address poverty, promote education, and improve healthcare systems, ensuring social welfare and reducing inequalities.

Collaboration between individuals, entities, and governments is essential to confront global challenges effectively. This can be achieved through public-private partnerships, where businesses work closely with governments to achieve common goals. Entities can engage in knowledge sharing, cooperate on initiatives, and align their objectives with government policies. Furthermore, governments can encourage and facilitate public participation by promoting citizen engagement, seeking public opinion, and incorporating diverse perspectives in policy-making processes.

In conclusion, addressing the responsibility of individuals, entities, and governments in tackling global issues requires a collective effort. Individuals need to take action in their personal capacity and demand sustainable practices from businesses. Entities must integrate sustainability into their operations and actively participate in community-building initiatives. Governments must formulate and implement policies that prioritize sustainability and social welfare. Together, these stakeholders can create a positive impact and work towards a sustainable and equitable future.

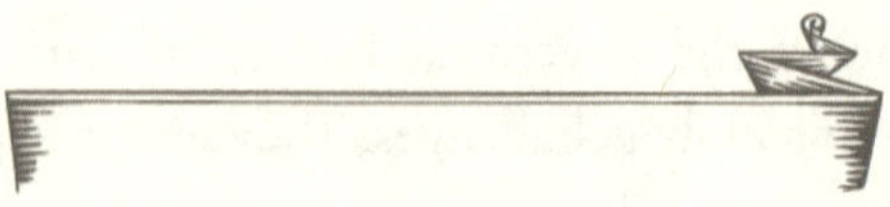

Part II: Connecting the Dots: Sea Level Rise

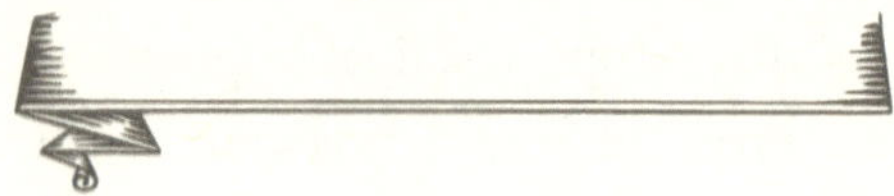

In Part II of this series, we delve deeper into the complexities and interconnectedness of sea level rise. We will explore several factors that contribute to the rise in sea level, as well as their implications for coastal communities around the world.

One of the key contributors to sea level rise is the melting of ice sheets, particularly in Antarctica and Greenland. These massive ice sheets hold enormous amounts of water that, once melted, flow into the ocean, causing sea levels to rise. This process is often accelerated by the warmer temperatures brought about by climate change. Not only does this result in higher sea levels, but it also leads to the destabilization of ice shelves that act as buttresses, holding back even greater volumes of ice. As these ice shelves collapse, more and more ice sheets are exposed to the warming ocean, further intensifying the cycle of sea level rise.

Another driver of sea level rise is thermal expansion. When the ocean absorbs heat from the atmosphere, it warms up and expands, occupying more volume. This expansion contributes significantly to the overall increase in sea levels. As global temperatures continue to rise, the ocean's capacity to absorb heat also increases, compounding the effects of thermal expansion.

In addition to these natural causes, human activities such as the extraction of groundwater and the damming of rivers can also contribute to sea level rise. When groundwater is extracted for various purposes, such as irrigation or drinking water supply, it eventually finds its way back to the ocean. Similarly, when rivers are dammed for hydroelectric power or water storage, the water that would have flowed into the ocean is retained, thereby raising sea levels.

The implications of sea level rise are significant and far-reaching. Coastal communities around the world are already feeling the effects, with increased flooding, erosion, and saltwater intrusion into freshwater aquifers. As sea levels continue to rise, these effects will only worsen, leading to the displacement of millions of people and the loss of valuable coastal ecosystems.

Coastal cities are particularly vulnerable to the impacts of sea level rise. Cities like Miami, New York, and Tokyo, are already investing considerable resources to protect themselves from rising waters. From constructing seawalls and flood barriers to elevating buildings and developing drainage systems, these cities are exploring various adaptation strategies to mitigate the risks associated with sea level rise. However, it is important to note that these measures are often costly and can only provide temporary relief, as sea levels are projected to continue rising for centuries to come.

The consequences of sea level rise extend beyond human habitats. Coastal ecosystems, such as mangroves, coral reefs, and saltmarshes, provide numerous ecological and economic benefits. These habitats serve as nurseries for fish and other marine species, protect the shoreline from erosion, and generate tourism revenue. As sea levels rise, these ecosystems will face increasing pressures, potentially leading to their degradation or even disappearance. This loss would have cascading impacts on biodiversity, fisheries, and the overall health of coastal ecosystems.

In conclusion, sea level rise is a complex and urgent challenge that requires global attention and concerted action. The melting of ice sheets, thermal expansion, and human activities all contribute to this phenomenon. The implications of rising sea levels are severe, affecting coastal communities and ecosystems worldwide. Adaptation and mitigation strategies are necessary to protect vulnerable areas and ensure the resilience of both human and natural systems. Continued research, cooperation, and sustainable practices are key to safeguarding our planet from the far-reaching impacts of sea level rise.

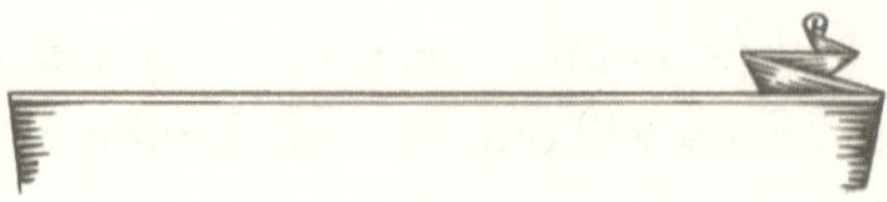

Chapter 5: Uncovering the Oceans' Role

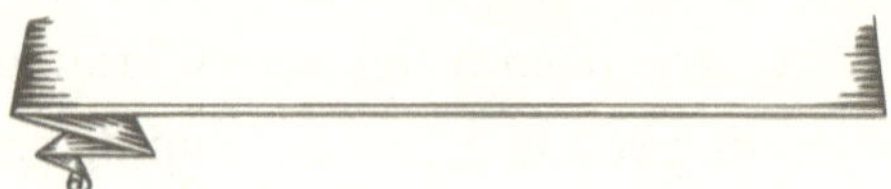

The vast expanses of the world's oceans have always fascinated humanity. From their mysterious depths to the life they support, the oceans play a vital role in the functioning of our planet. In this chapter, we delve into uncovering the oceans' role and the fascinating ways in which they impact various aspects of our lives.

The oceans act as a massive storage system for heat, absorbing and storing vast amounts of energy from the sun. This stored heat is responsible for regulating Earth's climate and weather patterns, making the oceans a critical player in the global climate system. Through a process known as thermohaline circulation, the oceans also redistribute heat throughout the planet, affecting temperature variations in different regions.

Not only do the oceans drive weather patterns on land, but they also influence the formation and intensification of hurricanes. Warm ocean waters provide the energy needed to fuel these massive storms, and factors such as sea surface temperature and current patterns determine their paths and strengths. By studying the oceans' role in hurricane development, scientists can better predict and understand these destructive natural phenomena.

In addition to climate regulation and storm formation, the oceans play a crucial role in absorbing carbon dioxide, a greenhouse gas responsible for global warming. The vast majority of the world's carbon dioxide emissions are absorbed by the oceans, serving as a natural carbon sink. However, this increased absorption has led to ocean acidification, a process that threatens marine life by reducing the availability of shell-building materials for organisms like corals and mollusks. Understanding the oceans' ability to absorb and mitigate carbon dioxide is vital for tackling climate change.

Moreover, the oceans are teeming with life, making them the most diverse ecosystems on the planet. From microscopic organisms to the largest creatures on Earth, the oceans provide habitats and nourishment for a myriad of species. Understanding the oceans' role in supporting biodiversity is not only important for ecological balance but also for the potential pharmaceutical and nutritional benefits that marine organisms can offer to human health and well-being.

Beyond their environmental significance, the oceans also hold vast economic value. They are a source of food, energy, and numerous raw materials, such as oil, gas, and minerals. Commercial fishing and aquaculture industries rely on the oceans for sustenance and profit, while offshore oil drilling and mineral exploration fuel economic growth. Understanding the oceans' role in supporting these industries is crucial for sustainable management and responsible use of these resources.

In this chapter, we have explored the multifaceted role of the oceans, from their influence on climate and weather patterns to their contributions to carbon absorption and marine life. Through scientific research and technological advancements, we continue to uncover the intricacies of the oceans' functioning and their interconnectedness with multiple aspects of the planet. Recognizing the importance of the oceans' role is essential for informed decision-making, conservation, and ensuring a sustainable future.

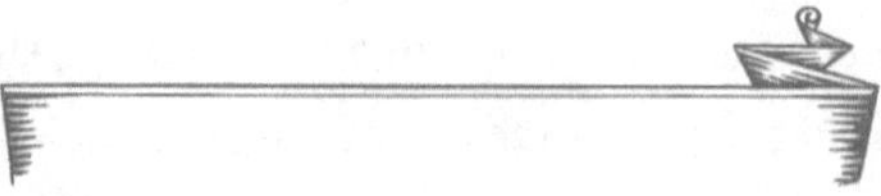

- Discovering the influential role of the oceans in climate change

The oceans play a crucial role in shaping Earth's climate, exerting a strong influence on both global and regional weather patterns. Their vastness, depth, and ability to absorb massive amounts of heat and carbon dioxide make them indispensable in understanding and predicting climate change.

Firstly, the oceans serve as a major heat reservoir. They absorb about 90 percent of the excess heat generated by human activities, thus mitigating the immediate impact of rising greenhouse gas emissions on the atmosphere and land. This absorption helps regulate global temperatures, preventing them from escalating too quickly. Without the oceans acting as a natural buffer, the Earth's surface would have warmed much faster than it has over the past century.

Moreover, the oceans act as engines of the global circulation system, redistributing heat across the planet. This circulation is driven by two main factors: temperature and salinity. Warm, tropical surface waters move towards the poles, where they release the absorbed heat into the atmosphere. In turn, the cooler water sinks to deeper layers, forming a global conveyor belt of heat distribution. This process, known as thermohaline circulation, contributes to regulating regional climates, including influencing monsoon patterns and the onset of El Niño and La Niña events.

Furthermore, oceans absorb vast quantities of carbon dioxide (CO_2), the primary greenhouse gas responsible for global warming. Through a process called oceanic uptake or carbon sequestration, CO_2 dissolves into seawater, where it undergoes complex chemical reactions. These reactions result in the formation of carbonic acid, which leads to ocean acidification. While this process helps to reduce the amount of CO_2 in the atmosphere, it has severe consequences for marine ecosystems, coral reefs, and shell-forming organisms.

RISING TIDES: THE CONNECTION BETWEEN GLOBAL WARMING AND SEA LEVEL RISE

The role of the oceans in climate change extends beyond heat absorption and carbon sequestration. Ocean currents and waves have a profound impact on coastal regions, influencing local climates, erosion, and precipitation patterns. For instance, the Gulf Stream, a powerful warm ocean current flowing from the Gulf of Mexico to the North Atlantic, keeps Western Europe much warmer than other regions at similar latitudes. Changes in ocean circulation patterns can disrupt these regional climates, leading to potentially dramatic shifts in weather patterns.

The interplay between the oceans and climate change also encompasses sea levels and ice melt. Melting ice caps and glaciers contribute to rising sea levels, putting coastal regions and low-lying islands at risk of inundation. As temperatures rise, polar ice caps and glaciers melt faster, largely due to the warming ocean waters. This meltwater adds to the volume of the ocean, further exacerbating the already increasing sea levels. Additionally, the melting of sea ice and ice shelves disrupts the delicate balance of ocean currents, affecting global weather systems and altering climate patterns.

In conclusion, the oceans are at the heart of Earth's climate system, exerting a critical influence on weather, atmospheric temperatures, sea levels, and carbon cycle dynamics. Understanding the complex interactions between the oceans and climate change is vital for accurately modeling and predicting future climate scenarios. Efforts to tackle global warming and mitigate the impacts of climate change must include measures to protect and sustain the health of our oceans, as they are essential for the health and well-being of our planet as a whole.

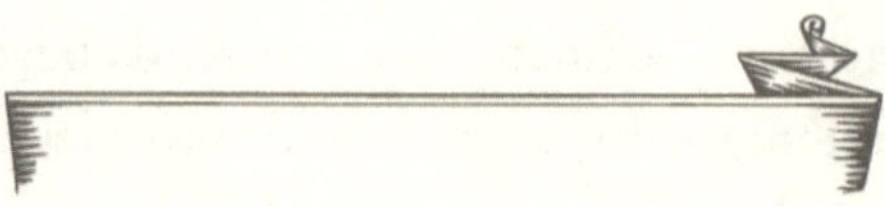

- Investigating thermal expansion, glacial melt, and their contributions to rising sea levels

Thermal expansion and glacial melt are two crucial factors contributing to the rise in global sea levels. Understanding these mechanisms is essential for predicting future sea level rise and its potential impacts on both coastal regions and global climate patterns. Various scientific studies have been undertaken to investigate and monitor these phenomena, shedding light on their complex mechanisms and their cumulative effect on sea level rise.

Thermal expansion refers to an increase in the volume of water as it absorbs heat energy. This occurs due to the fundamental nature of water molecules, which become energized, move more vigorously, and take up more space as they gain thermal energy. The more heat that is absorbed by water, the greater its expansion and subsequent increase in volume. As the average global temperature has been steadily rising due to anthropogenic greenhouse gas emissions, so too has the oceans' temperature. This rise in ocean temperatures has led to an expansion of water volume and subsequently contributed to the overall rise in sea levels.

Glacial melt, on the other hand, originates from mass loss in ice sheets and glaciers worldwide. Ice sheets, such as those covering Antarctica and Greenland, as well as mountain glaciers found in various regions globally, have experienced significant melting as a consequence of rising global temperatures. The warming climate leads to increased melting, with the resultant water making its way into the ocean, thereby contributing to sea level rise. Intensive research is being conducted to monitor ice sheet melt rates, understand their underlying processes, and predict their future behavior. The mass balance of glaciers and ice sheets is particularly crucial because a significant loss in ice mass has the potential to contribute to a substantial increase in sea level.

To investigate thermal expansion and glacial melt, various observational and modeling techniques are employed. Satellite altimetry, which uses satellite measurements of the Earth's gravitational field to monitor sea level changes, has been essential in providing global coverage and long-term records of sea level rise. These measurements contribute to our understanding of the rate at which sea levels are rising and identify regions experiencing relatively higher rates of change.

To complement satellite altimetry, tide gauge measurements are carried out along coastlines worldwide. These measurements provide local data on relative sea level, including details on local variations caused by factors like isostatic adjustments and tectonic movements. Combining satellite and tide gauge data allows scientists to assess regional sea level changes, including areas significantly impacted by thermal expansion or changes in ocean circulation patterns.

In addition to observations, numerical models are utilized to simulate and predict sea level rise. Climate models are typically based on various physical laws governing oceanic and atmospheric behavior, combined with computations of heat transfer, mass balance, and fluid dynamics. By running simulations with different input parameters, scientists can obtain projections of future sea level rise under various scenarios of greenhouse gas emission and global warming. Such models help inform policymakers and the scientific community about potential outcomes of different mitigation and adaptation strategies and provide vital evidence in ongoing climate change discussions.

Investigations into thermal expansion and glacial melt have brought about significant knowledge regarding sea level rise. These investigations reveal that the contributions from both processes are indisputable, emphasizing the need for concerted global efforts to mitigate climate change and reduce greenhouse gas emissions. Ultimately, understanding these mechanisms and their complex interactions is crucial to establishing effective strategies for managing the societal and environmental impacts of rising sea levels.

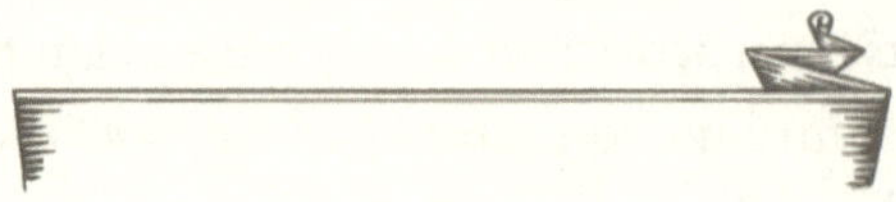

Chapter 6: Navigating the Effects: Coastal Areas

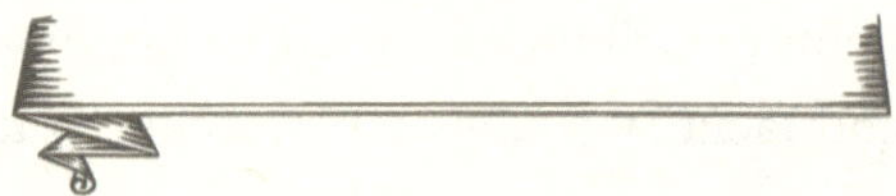

Coastal areas play a pivotal role in our planet's ecosystem, serving as a habitat for numerous species, providing a source of livelihood for millions of people, and offering stunning natural beauty. However, these regions are not impervious to the effects of human activity and climate change. In this chapter, we will explore the intricate dynamics of coastal areas and delve into the environmental consequences they face. Furthermore, we will discuss crucial strategies to navigate and mitigate these effects, ensuring the long-term sustainability and preservation of these precious landscapes.

1. Rising Sea Levels:

One of the most significant consequences of climate change is the gradual increase in sea levels. As global temperatures rise, glaciers melt, and thermal expansion occurs, coastal areas bear the brunt of the rising waters. Low-lying regions are especially vulnerable to flooding, endangering both human settlements and the delicate balance of coastal ecosystems. The relocation of communities, disrupted coastal fisheries, and loss of protective natural barriers are among the key consequences of rising sea levels.

2. Erosion and Sedimentation:

Coastlines constantly undergo erosion and sedimentation due to natural processes. Nonetheless, human activities, such as the construction of ports or the extraction of sand for industrial purposes, amplify these processes and exacerbate their effects. The loss of sand dunes, beach erosion, and the deterioration of coastal cliffs not only reduce the aesthetic value of coastal areas but also threaten human infrastructure and disrupt ecosystems' delicate dynamics.

3. Coral Bleaching and Ocean Acidification:

RISING TIDES: THE CONNECTION BETWEEN GLOBAL WARMING AND SEA LEVEL RISE

Coastal areas often boast vibrant coral reefs, which are hotspots of biodiversity and provide essential habitats for countless marine species. However, increasing ocean temperatures and acidity levels pose severe threats to these delicate ecosystems. Coral bleaching occurs when corals, stressed by warmer waters, expel their symbiotic algae, leading to their death. This, in turn, affects the entire food web and disturbs the delicate balance of coastal areas. Furthermore, the acidification of oceans negatively impacts shell-forming organisms, such as mollusks, impairing their ability to build protective shells.

4. Pollution and Eutrophication:

Coastal areas are prone to pollution, as human activities both on land and at sea result in the discharge of harmful substances. Chemical spills, industrial waste, and sewage discharge not only harm marine life but also degrade water quality, causing a ripple effect on both humans and ecosystems. Furthermore, excessive nutrient runoff from agricultural activities can lead to eutrophication, resulting in harmful algal blooms, oxygen depletion, and disruptions in nutrient cycling.

5. Overfishing and Habitat Destruction:

Coastal areas often witness intense fishing activity due to the abundance of marine resources. However, unsustainable fishing practices, such as overfishing, bottom trawling, and destructive fishing methods, jeopardize both target species and other non-targeted organisms. This leads to the depletion of fish populations, the collapse of food webs, and irreversible damage to essential habitats like seagrass beds and mangrove forests. Consequently, the loss of these critical ecosystems undermines coastal resilience, increases vulnerability to storms, and reduces the potential for carbon sequestration.

NAVIGATING THE EFFECTS of human activity and climate change in coastal areas necessitates a holistic approach that considers ecological, socioeconomic, and policy dimensions. Conservation efforts, restoration of critical habitats, sustainable fishing practices, and the implementation of pollution reduction measures are all pivotal in preserving the health and resilience of coastal areas. By recognizing the significance of these locations and addressing their vulnerabilities, we can strive towards a future where coastal areas continue to thrive while ensuring a sustainable and harmonious coexistence between humans and nature.

44

areas continue to thrive while ensuring a sustainable and harmonious coexistence between humans and nature.

- Highlighting the effects of rising sea levels on coastal ecosystems

The effects of rising sea levels have become a pressing issue in the world's coastal ecosystems. As global temperatures continue to increase due to human-driven climate change, the polar ice sheets and glaciers are melting at an alarming rate. This rapid melting has led to the rise in sea levels, which is now threatening the fragile balance of these unique ecosystems.

Coastal ecosystems are diverse habitats that consist of a wide range of species and are home to numerous important ecological processes. Mangrove forests, salt marshes, seagrass meadows, and coral reefs are some of the prominent examples of coastal ecosystems that exist across the globe. These ecosystems provide essential services such as carbon sequestration, shoreline protection, and habitat for various marine and terrestrial species. However, with the rising sea levels, these ecosystems are facing significant challenges.

One of the most immediate threats posed by rising sea levels is the submergence of low-lying coastal areas. Many coastal wetlands, such as salt marshes and mangrove forests, are naturally adapted to fluctuating water levels. However, the rate at which the sea levels are rising is unprecedented, and these ecosystems are not able to keep up with the increase. As a result, large areas of salt marshes and mangroves are being submerged, causing irreparable damage to these delicate environments.

Salt marshes play a crucial role in coastal ecosystems by acting as natural filtration systems, removing pollutants from the water and improving water quality. They also act as buffers during storms, absorbing wave energy and protecting the surrounding coastal areas from erosion. However, with rising sea levels, the frequency and intensity of these storms are also increasing, posing a double threat to salt marshes. Flooding from storm surges combined with the steady rise in sea levels can lead to the loss of these valuable marshes, with severe consequences for coastal communities.

Similarly, mangrove forests are vital coastal ecosystems that provide a unique habitat for numerous marine and terrestrial species. These forests act as carbon sinks, capturing and storing vast amounts of carbon dioxide, thus mitigating the impacts of climate change. However, rising sea levels are causing saltwater intrusion into the mangroves, making it difficult for them to thrive. This leads to significant losses in terms of biodiversity and ecosystem services provided by the mangroves.

Seagrass meadows and coral reefs are also vulnerable to rising sea levels. Seagrasses are key habitats for many species, acting as nurseries and providing food for various marine animals. Moreover, they also contribute to carbon sequestration. However, increased water depth caused by rising sea levels can affect the growth and survival of seagrass meadows, resulting in their decline. Coral reefs, on the other hand, depend heavily on sunlight and shallow water. As sea levels rise, many coral reefs are becoming stressed and are more prone to bleaching events and subsequent mortality.

In addition to these direct impacts, rising sea levels also result in shoreline erosion, leading to the loss of important coastal habitats and threatening human settlements. As the ocean encroaches further inland, coastal communities face a heightened risk of flooding and increased vulnerability to coastal hazards such as storm surges and hurricanes.

To mitigate the effects of rising sea levels on coastal ecosystems, it is imperative to take immediate action. Coastal management strategies such as beach nourishment, coastal reforestation, and the creation of artificial reefs can help protect vulnerable coastal habitats. Implementing sustainable land-use practices, reducing greenhouse gas emissions, and promoting climate change adaptation measures can play a crucial role in preserving coastal ecosystems in the face of rising sea levels.

The impact of rising sea levels on coastal ecosystems cannot be overstated. From the loss of vital habitat for marine and terrestrial species to increased vulnerability of coastal communities, the consequences are far-reaching. Recognizing the importance of these ecosystems and working towards their preservation is crucial for the long-term sustainability of both the environment and human populations that depend on them.

- Discussing coastal erosion, loss of habitat, and human displacement

Coastal erosion is an ongoing process that affects the fragile ecosystems and habitats along coastlines worldwide. It occurs due to a combination of natural factors such as waves, tides, and wind, as well as human activities such as sand mining, urbanization, and climate change. As a result, many coastal regions are losing their natural defenses against erosion, leading to the loss of habitats and even human displacement.

One of the primary consequences of coastal erosion is the destruction of valuable wildlife habitats. Many coastal areas serve as nesting sites for migratory birds, breeding grounds for fish, and habitats for numerous other species, including turtles, marine mammals, and various plant species. These ecosystems are critically important for the biodiversity of the planet, as they support a wide range of species and are interconnected with other ecosystems such as coral reefs and mangrove forests. Unfortunately, as erosion intensifies, these habitats are gradually disappearing, leaving the wildlife that depends on them vulnerable and at risk of extinction.

Moreover, coastal erosion also impacts human settlements located near the coast. For centuries, people have been drawn to coastal areas due to their scenic beauty, availability of natural resources, and accessibility for trade purposes. However, as erosion takes hold, it erodes beachfronts, cliffs, and dunes, threatening the infrastructure, property, and even lives of coastal communities. Villages, towns, and cities built near the shoreline are increasingly prone to flooding, storm surges, and saltwater intrusion, which pose significant risks to the safety and well-being of the inhabitants.

In addition to the direct physical impacts, coastal erosion often leads to the displacement of human populations. As erosion intensifies, communities are forced to abandon their homes and seek alternative locations to settle. This leads to a host of social, economic, and logistical challenges for both the displaced populations and the regions they move to. Displaced individuals may find themselves in unfamiliar environments, face housing issues, joblessness, and even conflicts with their new neighbors. These consequences can be particularly severe in less developed countries, where resources and infrastructure for relocation are often insufficient, exacerbating the vulnerability of already marginalized communities.

Climate change is a significant contributing factor to the acceleration of coastal erosion in recent years. Rising sea levels, attributed to global warming and the melting of polar ice caps, exacerbate the impacts of erosion. As sea levels rise, the power and reach of waves increase, leading to more severe and widespread erosion. Furthermore, climate change also influences weather patterns, potentially increasing the frequency and intensity of storms and hurricanes, which cause immediate and substantial damage to coastal areas.

Addressing coastal erosion and its associated consequences requires a multifaceted approach. Firstly, coastal management strategies must be employed to preserve and restore natural coastal defenses such as dunes, wetlands, and mangrove forests. These ecosystems act as effective barriers against erosion, absorbing wave energy and stabilizing sediments. Deforestation and land development near the coast should also be regulated to minimize the removal of natural buffers.

Furthermore, governments and communities at risk should invest in the construction of engineered structures such as sea walls, groynes, and breakwaters. While these structures can help mitigate erosion and protect infrastructure in the short term, they should be considered as part of a broader solution, as they can have unintended consequences on neighboring coastlines and ecosystems.

Long-term measures to combat coastal erosion necessitate proactive climate change mitigation and adaptation strategies. Reducing greenhouse gas emissions, transitioning to renewable energy sources, and promoting sustainable development practices are imperative at the global level. Locally, municipalities should implement comprehensive planning processes that factor in climate change projections when determining coastal land use, infrastructure construction, and community development.

Preserving and securing the world's coastline is not only vital for the protection of diverse wildlife habitats but also for the well-being and safety of coastal communities. The challenges posed by coastal erosion require a combination of ecological, engineering, and policy-based solutions, coupled with community engagement and support from governments and international organizations. Only through collective efforts can we ensure the sustainability and resilience of our coastal regions for generations to come.

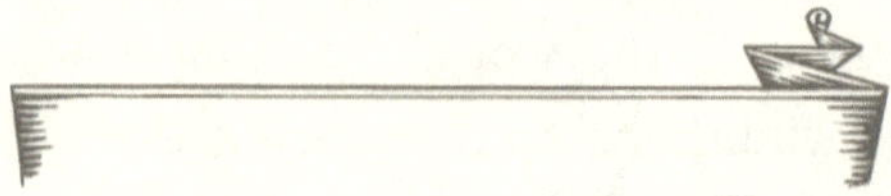

Chapter 7: The Cost of Melting Ice: Polar Regions

The polar regions, comprising the Arctic and Antarctic, are facing a crisis caused by the melting of their pristine ice. This chapter explores the diverse consequences of this ice loss, including ecological disruptions, rising sea levels, and the socio-economic impact on indigenous communities. The complexity and importance of preserving these polar environments are highlighted, urging us to reconsider our actions and seek sustainable solutions.

1. Ecological Disruptions:

The melting ice in polar regions has led to significant ecological disruptions. Polar bears, iconic symbols of the Arctic, are losing their natural habitat as the sea ice progressively disappears. Their hunting grounds for seals are diminishing, affecting both their ability to reproduce and their overall population numbers. Other Arctic species, such as reindeer, arctic foxes, and walrus, are also facing considerable challenges due to the melting ice, ultimately disrupting the delicate balance of polar ecosystems.

2. Rising Sea Levels:

Another critical consequence of melting ice in polar regions is the rising sea levels worldwide. The ice sheets of Antarctica and Greenland hold vast quantities of water, and as they melt, this water flows into the oceans, causing sea levels to rise. Low-lying coastal areas around the world are particularly vulnerable to the impacts of this rise, facing increased risks of flooding and storm surges. Furthermore, small island nations are at a heightened threat of disappearing entirely as their homes are swallowed by the encroaching waters.

3. Socio-Economic Impact:

The melting ice also brings significant socio-economic impacts to indigenous communities living in the polar regions. Many indigenous cultures depend on hunting and fishing as their primary sources of sustenance. With the melting ice altering traditional hunting routes and compromising the availability of marine mammals, these communities face an uncertain future. Their cultural identity, deeply connected to these unique environments, is eroding along with the melting ice. Additionally, tourism in these regions is also suffering as iconic landscapes and wildlife diminish, impacting local economies.

4. Climate Change Feedback Loops:

Melting ice in the polar regions contributes to an amplified feedback loop of climate change. As the ice melts, it exposes darker ocean surfaces and unveils previously hidden permafrost, both of which absorb more sunlight and heat, accelerating the warming of the region. This intensification of climate change affects not only the polar regions but also influences weather patterns and global climate stability.

THE MELTING ICE IN polar regions is immensely costly, both ecologically and socio-economically. This chapter has emphasized the complex web of consequences resulting from this ice loss, ranging from disrupted ecosystems to rising sea levels and increased vulnerability for indigenous communities. It is crucial for humanity to recognize and address the looming crisis, finding sustainable solutions to preserve these polar environments. By doing so, we can minimize the devastating costs of melting ice and ensure a more harmonized future for our planet.

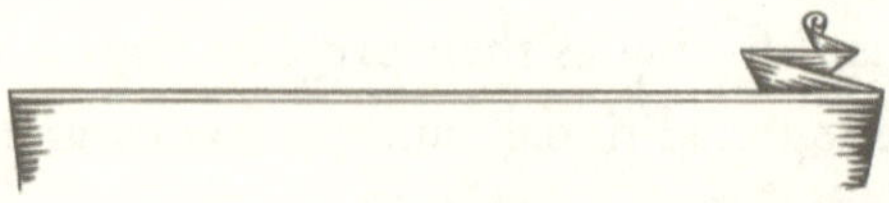

- Focusing on the perilous effects climate change has on the Arctic and Antarctic regions

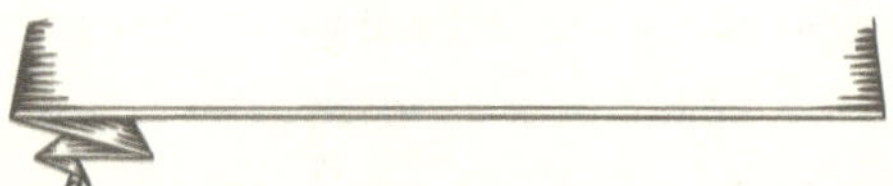

Climate change is a global issue that affects various regions across the planet. However, no areas are as vulnerable and visibly impacted by this phenomenon as the Arctic and Antarctic regions. These polar environments serve as crucial indicators of global climate fluctuations, making them vital to our understanding of the Earth's changing climate and its devastating consequences.

The Arctic, comprised of the Arctic Ocean and surrounding land masses, has experienced dramatic changes in recent years. Rising temperatures have caused a rapid decline in sea ice, with the Arctic's summer sea ice extent decreasing by nearly 13% per decade. This loss has grave implications for the region's fragile ecosystem, as it threatens the habitat and survival of iconic ice-dependent species like polar bears, seals, and walruses. Additionally, declining sea ice allows for increased shipping activities and resource extraction, which poses significant risks to marine life and indigenous communities.

Melting permafrost is another pressing issue challenging the Arctic's stability. Permafrost is permanently frozen ground that occurs in nearly half of the region's land area. As temperatures rise, this frozen ground thaws, releasing vast amounts of trapped greenhouse gases into the atmosphere. The release of carbon dioxide and methane stored in the permafrost exacerbates global warming, contributing to a dangerous feedback loop.

While the Arctic witnesses these rapid changes, the Antarctic, located at the southernmost part of Earth, is also experiencing dire consequences. Although the Antarctic is colder and less populated than the Arctic, it faces serious challenges brought about by climate change. The collapse of ice shelves and unstable glaciers in the region have resulted in rising sea levels, a threat that could affect millions of people living in coastal areas around the world.

The West Antarctic Ice Sheet, in particular, is near a catastrophic tipping point known as the "point of no return." If the ice sheet completely melted, it could lead to an estimated three-meter rise in sea level. This, combined with the ongoing disintegration of the Antarctic Peninsula's ice shelves, poses significant dangers to low-lying islands and densely populated coastal regions, placing countless lives and infrastructure at risk.

A warming climate has also disrupted the delicate balance of Antarctic ecosystems. Krill, tiny shrimp-like organisms, serve as a linchpin of the food chain in the Southern Ocean, supporting several species of whales, penguins, and seals. However, as sea temperatures increase and sea ice declines, krill populations are at risk of significant declines, leading to ripple effects throughout the entire ecosystem.

Moreover, the climate crisis threatens the Antarctic's remarkable biodiversity and endemic species.

Species like the Emperor penguin, Weddell seal, and Adélie penguin have specific habitat requirements that depend on stable sea ice conditions, which are rapidly diminishing. As a result, these unique species face the risk of population declines and potential extinctions, greatly diminishing Earth's biological diversity.

It is clear that climate change is rapidly transforming both the Arctic and Antarctic regions, with grave consequences for the environment and its inhabitants. Urgent action is required on a global scale to reduce greenhouse gas emissions, transition towards renewable energy sources, and protect these delicate ecosystems. By doing so, we can strive towards mitigating the impending dangers that climate change presents to these crucial parts of our world.

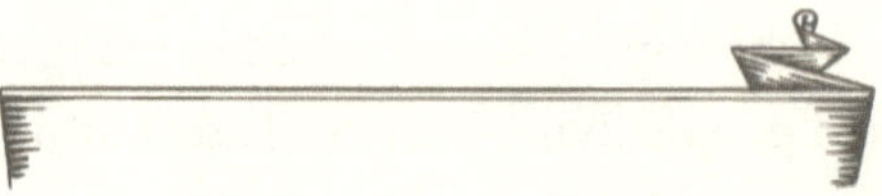

- Discussing the melting ice sheets, the impact on ecosystems, and implications for global sea levels

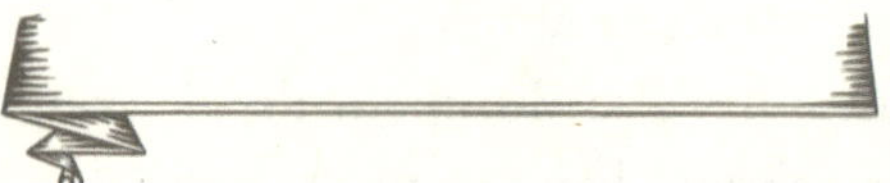

Melting ice sheets are a significant consequence of climate change and have been a growing concern in recent decades. These massive ice structures, found predominantly in Greenland and Antarctica, hold immense volumes of frozen water. As temperatures rise, the integrity of these ice sheets is compromised, causing them to gradually melt and ultimately contribute to the rise in global sea levels. This rapid melting has crucial implications not only for the physical landscape but also for ecosystems and human society.

The immediate impact of melting ice sheets is felt locally in the polar regions, where the most substantial ice sheets are found. As these ice masses melt, an enormous amount of freshwater pours into the oceans. This influx of freshwater alters the composition and salinity levels of seawater. Fragile marine ecosystems, especially all species that are specifically adapted to thrive in a specific range of salinity, get disrupted and struggle to adapt to the rapid changes. Many oceanic organisms rely on particular temperature and salinity gradients for their survival, and if these balances are disrupted, potentially irreparable damage can occur to these delicate ecosystems.

Moreover, melting ice sheets also have far-reaching effects on the global climate, exacerbating climate change itself. The layer of ice on land reflects sunlight back into space, efficiently reducing the amount of heat energy absorbed by the Earth's surface. As this ice melts, it exposes the darker ground or seawater beneath, which then absorbs more sunlight and subsequently amplifies warming. This feedback loop can lead to a further increase in global temperatures and accelerate the melting process.

RISING TIDES: THE CONNECTION BETWEEN GLOBAL WARMING AND SEA LEVEL RISE

The enhanced rate of ice loss also has severe implications for global sea levels. Studies have shown that if all the ice sheets in Greenland or Antarctica were to melt entirely, it would lead to an astonishing sea-level rise of 6 to 7 meters (20 to 23 feet) respectively. Although such a catastrophic event is unlikely to occur in the short term, even a relatively modest increase in sea levels can result in a plethora of dramatic consequences.

Low-lying coastal regions, particularly heavily populated ones, are at the highest risk, facing increased vulnerability to tidal floods, surge events, and storm surges. Many of the world's major cities are situated near coastlines and would be significantly impacted. Infrastructure, urban settlements, and valuable ecosystems located in these areas are threatened by flooding, leading to an increase in economic and humanitarian burdens. Additionally, saltwater intrusion further inland would contaminate freshwater resources, leading to water scarcity and risking the quality of drinking water supplies.

The impacts of melting ice sheets are not limited to coastal regions, as changes in global climate patterns induced by melting ice can have broader consequences. Melting ice affects ocean currents, which play a critical role in distributing heat and nutrients around the world. Altering these currents can disrupt major climate systems such as the Gulf Stream, which transports warm water from the tropics to higher latitudes, affecting both regional and global weather patterns. This disruption can result in more frequent and intense weather events, such as hurricanes, heatwaves, or prolonged droughts, adding to the challenges already faced by communities worldwide.

In conclusion, the melting of ice sheets due to climate change carries vast implications for ecosystems, sea levels, and global climate patterns. The accelerated ice loss contributes to rising sea levels, posing a threat to coastal areas and their associated infrastructure, human populations, and fragile ecosystems. Additionally, changes in salinity and temperature gradients disrupt marine ecosystems, further disrupting the delicate balance of nature. As the impacts of melting ice sheets continue to unfold, urgent action is required to mitigate climate change and its cascading consequences.

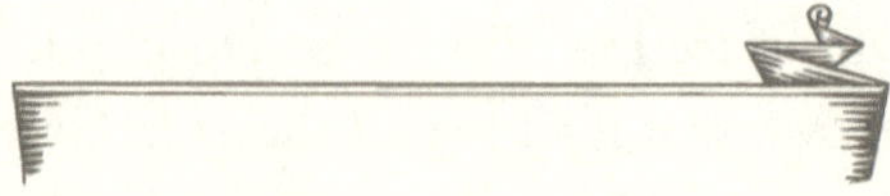

Chapter 8: Island Nations and Beyond

In this chapter, we delve into the fascinating world of island nations and explore how they have both unique challenges and inherent beauty. From the lush tropical paradises like the Maldives and Fiji to the remote and isolated Easter Island and Madagascar, these islands offer insights into the intricate relationship between human society and the natural world.

First, we take a closer look at the Maldives, a small archipelago in the Indian Ocean known for its breathtaking beaches, turquoise waters, and luxury resorts. However, underneath its idyllic surface lies the harsh reality of environmental challenges. Rising sea levels threaten the very existence of these low-lying islands, and the Maldivian government has been at the forefront of efforts to combat climate change and preserve their nation. We explore the innovative strategies employed by the Maldivians, from building floating gardens to implementing sustainable tourism practices, in their battle against environmental threats.

Moving on to another island nation, Fiji, we discover a land of vibrant culture and natural wonders. The Fijians exemplify the concept of "island time," living in harmony with the rhythms of nature and embracing a slower pace of life. However, globalization and tourism have brought new challenges to their way of living. We examine the impacts of mass tourism on the indigenous Fijian communities, exploring how they navigate between economic opportunities and concerns about cultural preservation.

Easter Island serves as a prime example of both the magnificence and fragility of island ecosystems. This remote and enigmatic island is famous for its massive stone statues, known as moai, which have puzzled researchers for centuries. We delve into the mysteries surrounding the creation of these statues and the decline of the island's civilization, examining the role of overexploitation of resources and environmental degradation in their downfall. Easter Island stands as a cautionary tale, reminding us of the delicate balance between human society and the natural environment.

Another extraordinary island nation is Madagascar, a biodiversity hotspot teeming with unique flora and fauna found nowhere else on Earth. We explore the intertwining relationship between the Malagasy people and their natural surroundings, from traditional practices of using medicinal plants to the challenges posed by deforestation and illegal wildlife trade. Madagascar highlights the significance of sustainable development and the preservation of biological diversity in island nations.

Beyond these specific island nations, we examine the broader themes and issues that encompass island studies. We delve into the concept of "insularity," the notion that living on an island shapes not only one's physical reality but also their social, cultural, and economic context. We also explore the effects of colonization and decolonization on island nations, discussing how these historical processes have influenced their trajectory.

Overall, this chapter offers a comprehensive and thought-provoking exploration of island nations, focusing on their unique challenges and captivating beauty. Whether it is the environmental threats faced by the Maldives, the delicate balance between tourism and culture in Fiji, the enigmatic history of Easter Island, or the biodiversity of Madagascar, island nations reveal the complex interplay between human societies and the natural world. Through this lens, we gain a deeper understanding of the intricate relationship between people and their island homes, and the importance of sustainable practices for their future survival.

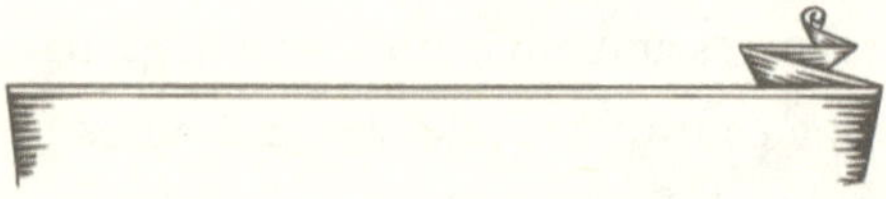

- Examining the vulnerability of low-lying areas and islands to sea level rise

Examining the vulnerability of low-lying areas and islands to sea level rise is of critical importance as climate change continues to alter our natural environment. The rising sea levels pose a significant threat to coastal regions, making it necessary to understand and assess the vulnerability of these areas and islands to formulate suitable adaptation and mitigation strategies.

Low-lying areas and islands are particularly susceptible to sea level rise due to their inherent geographical characteristics. These areas tend to have minimal elevation and rely heavily on the stability of coastal ecosystems for protection. When sea levels rise, the risk of coastal erosion and flooding increases dramatically, leading to severe consequences for human settlements, infrastructure, agriculture, and ecosystems.

One significant factor contributing to the vulnerability of low-lying areas is the subsidence or relative sea-level rise. Subsidence occurs when the land sinks, either naturally or due to factors like human activities, excessive groundwater extraction, or tectonic movements. Subsidence can amplify the sea level rise experienced in an area, exacerbating the vulnerability of low-lying regions and islands.

Another crucial aspect in assessing vulnerability is understanding the power of storm surges. Storm surges are elevated water levels caused by the combination of high tides, strong winds, and low atmospheric pressure during severe weather events such as hurricanes or cyclones. Low-lying areas and islands are highly exposed to the destructive force of storm surges due to their proximity to the coast and lack of natural protective barriers, such as coral reefs or wetlands.

RISING TIDES: THE CONNECTION BETWEEN GLOBAL WARMING AND SEA LEVEL RISE

The social and economic factors play a crucial role in exacerbating the vulnerability of these areas. Many low-lying regions and islands are densely populated and highly developed, characterized by tourism, fisheries, and various industries. As sea levels rise, these social and economic systems face grave threats. The displacement of communities, loss of livelihoods, damage to critical infrastructure, and increased financial burdens are just a few examples of the potential consequences. Moreover, marginalized communities and developing countries often bear a heavier burden due to their limited resources and lack of adaptive capacity.

It is essential to analyze and monitor these vulnerable areas and islands to develop effective adaptation strategies. This includes conducting detailed assessments of the potential impacts of sea level rise, identifying priority areas for potential interventions, and evaluating the feasibility of different adaptation options.

Nature-based solutions can play a significant role in reducing the vulnerability of low-lying areas and islands. Protecting and restoring coastal ecosystems, such as wetlands, coral reefs, and dunes, can act as a natural buffer against rising sea levels, reducing wave energy, and preventing coastal erosion. Additionally, implementing sustainable land management practices, regulating groundwater extraction, and considering spatial planning measures can contribute to enhancing the resilience of these areas.

Collaboration between different stakeholders is crucial for addressing the vulnerability of low-lying areas. Governments, research institutions, community organizations, and private sectors should work together to gather and share data, develop and implement adaptation plans, and provide support to affected communities. International cooperation and policy agreements are also vital to address the transboundary nature of sea level rise and effectively protect vulnerable areas.

In conclusion, examining and mitigating the vulnerability of low-lying areas and islands to sea level rise is a complex and urgent task. By understanding the specific challenges faced by these regions, implementing suitable adaptation and mitigation strategies, and fostering collaboration between stakeholders, it is possible to reduce the potential impacts and enhance the resilience of these coastal areas to a changing climate.

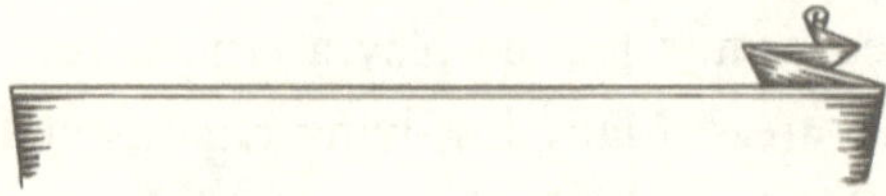

- Case studies highlighting the challenges faced by nations and communities threatened with inundation

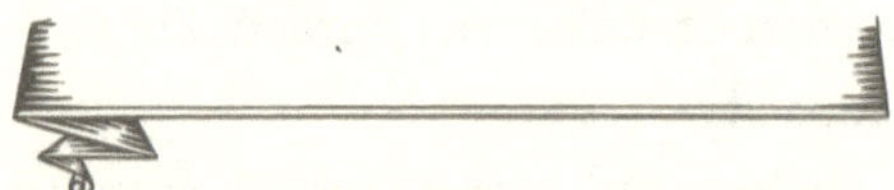

Case Studies on Inundation Threats: Unveiling Challenges Faced by Nations and Communities

INUNDATION, OR THE threat of being submerged underwater permanently, poses significant challenges for nations and communities worldwide. Rising sea levels and catastrophic events such as hurricanes are exacerbating this threat, requiring nations and communities to tackle various challenges. This article presents a collection of case studies that delve into the complexities surrounding the inundation threat, exploring different challenges faced by nations and communities.

Case Study 1: The Maldives- Rising Sea Levels

The Maldives is a low-lying archipelago nation situated in the Indian Ocean. One of the significant challenges faced by the Maldives is its vulnerability to rising sea levels, resulting from climate change. As the highest point in the Maldives is only 2.4 meters above sea level, the entire nation is prone to inundation. The government is grappling with issues related to infrastructure protection, relocation, and sustainable development in the face of this imminent threat.

Case Study 2: New Orleans, USA- Hurricane Katrina

Hurricane Katrina, which struck New Orleans in 2005, revealed the multiple challenges faced by the city and its inhabitants. The impact of the hurricane shattered levees and led to extensive flooding, causing massive destruction. The case study highlights the challenges of ensuring early warning systems, effective disaster response management, post-disaster recovery, and equitable rebuilding efforts to address the needs of diverse communities and prevent future inundation risks.

Case Study 3: Venice, Italy- Acqua Alta

Venice is widely known for its struggles with "acqua alta," or high waters. Located on a lagoon surrounded by the Adriatic Sea, the city is subject to periodic flooding events. Challenges faced by Venice include the need for innovative urban planning, sustainable engineering solutions, and protective measures. The case study showcases the role of community engagement, infrastructure upgrades, and research efforts tackling the multifaceted challenges of inundation threat in a historical and culturally significant city.

Case Study 4: Pacific Island Nations- Climate Change Impact

Numerous Pacific Island nations, including Kiribati and Tuvalu, are under constant threat due to inundation caused by climate change-induced rising sea levels. These low-lying islands face unprecedented challenges, with the potential loss of land, displacement of communities, and survival-related issues. The case study addresses the challenges associated with relocation, adaptation strategies, advocacy efforts at global platforms, and preservation of cultural identities while combating the threat of inundation.

INUNDATION THREATS inflict tremendous challenges on nations and communities, necessitating innovative and adaptive approaches. The case studies presented in this article illuminate some of the intricate hurdles faced by nations and communities around the world grappling with inundation threats. Understanding these challenges empowers decision-makers, planners, and communities to develop robust strategies that encompass infrastructure upgrades, disaster management plans, community engagement, and sustainable development practices to mitigate the risks posed by inundation.

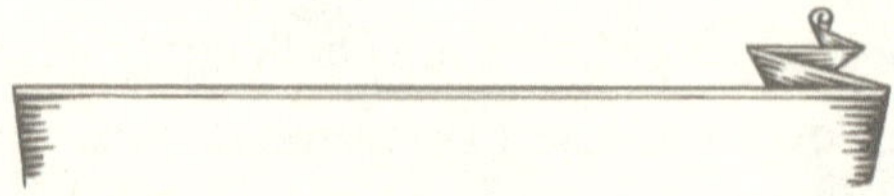

Part III: Seeking Solutions: Mitigation and Adaptation

In this part of the essay, we will be exploring the various approaches to combating and adapting to climate change – mitigation and adaptation. These two key strategies are essential for addressing the pressing issue of climate change and creating a sustainable future for our planet. While mitigation aims to reduce greenhouse gas emissions to prevent further warming, adaptation focuses on increasing resilience to the changes that are already occurring. By examining both approaches, we can gain a comprehensive understanding of the challenges and potential solutions in combatting climate change effectively.

Mitigation:

Mitigation strategies primarily aim to reduce greenhouse gas emissions, which are responsible for the majority of global warming. This approach recognizes the importance of limiting the release of harmful gases into the atmosphere. One of the primary ways to achieve mitigation is by transitioning from fossil fuels to renewable energy sources such as solar, wind, and hydroelectric power. Not only do renewable energy sources produce less greenhouse gas emissions, but they are also a more sustainable long-term solution for our energy needs.

Additionally, another key aspect of mitigation is improving energy efficiency. Enhancing energy efficiency in industries, buildings, and transportation can significantly reduce energy consumption and thus the emission of greenhouse gases. This can be achieved through the adoption of energy-saving technological advancements, stricter energy consumption regulations, and incentivizing the use of energy-efficient machinery and appliances.

Moreover, another crucial component of mitigation is forest restoration and conservation. Forests act as carbon sinks, removing carbon dioxide from the atmosphere and storing it in trees and soil. Preserving and expanding forests, especially in areas prone to deforestation, can greatly contribute to reducing overall greenhouse gas emissions. It is also important to address large-scale industrial agricultural practices that contribute to deforestation, such as palm oil production, and implement sustainable farming methods.

Adaptation:

While mitigation focuses on preventing climate change, adaptation aims to minimize the harmful impacts of climate change that are already happening. It recognizes that some degree of global warming is inevitable, and therefore, society must adapt to these changes. Adaptation strategies involve adjusting various aspects of human living to cope with the altered environmental conditions.

One of the key areas of adaptation is enhancing infrastructure resilience. This includes implementing water management systems capable of withstanding increased flooding and ensuring the stability of critical infrastructure such as power plants and transportation systems in the face of extreme weather events. Proper urban planning that accounts for rising sea levels and intensifying storms is crucial for minimizing the damage caused by climate change and safeguarding human lives.

Moreover, agricultural adaptation is of utmost importance since changing weather patterns greatly impact food production. Implementing drought-tolerant crops, improving irrigation techniques, and promoting sustainable farming methods can all contribute to increased resilience in agricultural systems. Similarly, protecting and rehabilitating coastal ecosystems, such as wetlands and reefs, not only provides habitats but also helps in protecting coastal communities from storm surges and erosion.

Planning for climate-induced health risks is also an essential aspect of adaptation. Rising temperatures increase the incidence of heatwaves, droughts, and the spread of vector-borne diseases. Reducing urban heat islands, providing cooling centers during heatwaves, and implementing effective public health strategies can help mitigate the health risks associated with climate change.

MITIGATION AND ADAPTATION are two distinct approaches to combating climate change and working towards a sustainable future. While mitigation strives to reduce greenhouse gas emissions and prevent further warming, adaptation focuses on minimizing the adverse impacts that are already inevitable. Both strategies are essential to address the multifaceted challenges of climate change effectively.

By transitioning to renewable energy sources, improving energy efficiency, and preserving forests, we can significantly reduce greenhouse gas emissions and mitigate climate change. Simultaneously, by enhancing infrastructure resilience, implementing sustainable agricultural practices, and planning for climate-induced health risks, we can adapt to the changes brought about by global warming and reduce the potential for disruption.

It is crucial that individuals, governments, and international bodies collaborate on both fronts to ensure a collective effort in combating climate change. Through timely and effective mitigation and adaptation strategies, we can create a sustainable future for generations to come, where the impacts of climate change are minimized, and our planet thrives.

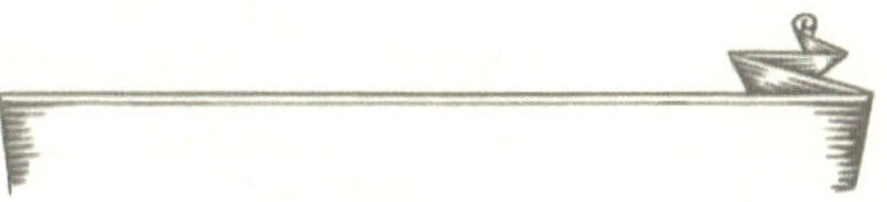

Chapter 9: Mitigating the Impacts of Global Warming

Global warming, caused primarily by human activities, is considered one of the most critical challenges humanity faces today. Its impacts on the environment, economy, and society are extensive and often devastating. However, as daunting as it may seem, there are steps that can be taken to mitigate the impacts of global warming. In this chapter, we will delve into various strategies and ways individuals, communities, and governments can contribute to lessening the effects of this imminent crisis.

1. Transition to Renewable Energy Sources:

One of the biggest contributors to global warming is the burning of fossil fuels for energy production. To tackle this issue, a crucial step is transitioning from fossil fuels to renewable energy sources such as solar, wind, hydro, and geothermal power. Governments and businesses must invest in green energy technologies, creating incentives for their adoption by individuals and industries.

2. Energy Efficiency and Conservation:

Besides switching to renewables, reducing energy consumption and increasing efficiency play a significant role in mitigating global warming. Initiatives like promoting energy-efficient appliances, buildings, and transportation systems are key. Implementing dynamic pricing mechanisms, incentives, and awareness campaigns can encourage individuals to conserve energy and thus reduce greenhouse gas emissions.

3. Sustainable Agriculture Practices:

Agriculture is responsible for a substantial portion of global greenhouse gas emissions. Shifting towards sustainable agricultural practices can considerably reduce these emissions while also adapting to the changing climate. Practices like precision farming, organic farming, and agroforestry can improve soil health, reduce the use of chemical fertilizers, and sequester carbon. Furthermore, promoting a plant-based diet can lower greenhouse gas emissions associated with the farming of livestock.

4. Reforestation and Forest Conservation:

Forests act as crucial carbon sinks, absorbing carbon dioxide and reducing atmospheric concentrations of greenhouse gases. Hence, efforts to proactively restore and protect forests are essential for mitigating global warming. Governments should support reforestation initiatives, implement policies against deforestation, and promote sustainable forestry practices. Organizations and individuals can also contribute through afforestation programs and protected area establishment.

5. Enhancing Transportation:

The transportation sector is a significant contributor to global warming. To mitigate its impacts, a transition to low-carbon alternatives is imperative. Promoting electric vehicles, investing in public transport systems, and encouraging cycling and walking can help reduce greenhouse gas emissions. Additionally, further research and development of cleaner aviation and shipping technologies can minimize their carbon footprints.

6. Circular Economy and Waste Management:

Waste generation and inefficient resource use contribute to greenhouse gas emissions. Adopting a circular economy approach that focuses on reducing, reusing, and recycling resources can significantly mitigate global warming effects. Moreover, transitioning to sustainable waste management practices, such as composting and waste-to-energy technologies, can further cut emissions while promoting resource conservation.

7. Education and Awareness:

Increasing public awareness about the impacts of global warming is crucial for its mitigation. Governments and institutions should make education and awareness campaigns a priority, educating individuals about the risks, causes, and consequences of climate change. Building literacy on sustainable lifestyle choices, clean energy practices, and responsible consumption can empower individuals to actively contribute towards reducing global warming impacts.

8. International Cooperation and Policy Frameworks:

RISING TIDES: THE CONNECTION BETWEEN GLOBAL WARMING AND SEA LEVEL RISE

Global warming is a global challenge that requires global solutions. International collaboration and agreements such as the Paris Agreement establish a framework to mitigate greenhouse gas emissions and limit temperature rise. It is important for governments to honor their commitments and work towards ambitious emission reduction targets. By sharing knowledge, resources, and best practices, countries can collectively address global warming and its impacts effectively.

MITIGATING THE IMPACTS of global warming necessitates comprehensive and interconnected actions at all levels. Individual choices, community initiatives, and government policies must align to transition to a more sustainable and resilient future. By adopting cleaner energy sources, implementing eco-friendly practices in various sectors, and fostering awareness and cooperation, humanity can take meaningful steps towards lessening the consequences of global warming and ensuring a better world for future generations.

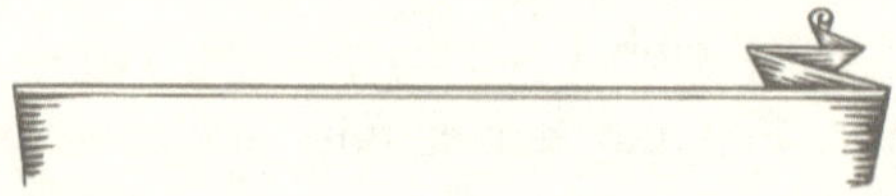

- Discussing potential strategies to mitigate global warming and subsequent sea level rise

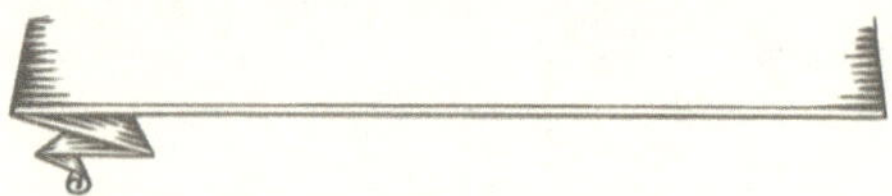

Discussing Potential Strategies to Mitigate Global Warming and Subsequent Sea Level Rise

GLOBAL WARMING, PRIMARILY caused by human activities, has become a pressing concern, leading to detrimental effects such as sea level rise, extreme weather events, and ecosystem disruptions. To mitigate these impacts, aggressive measures must be implemented on a global scale. This article delves into potential strategies that can be pursued to combat global warming and subsequent sea level rise, aiming to stabilize the climate while preserving coastal regions.

1. Transitioning to Renewable Energy Sources:

Phasing out fossil fuels and transitioning to renewable energy sources is crucial to mitigate global warming. Governments and international bodies can incentivize the development and deployment of solar, wind, hydroelectric, and geothermal power sources. Encouraging research and development in battery storage and green hydrogen technologies will also aid in the transition.

2. Energy Efficiency and Industrial Practices:

Promoting energy efficiency in various sectors, including manufacturing, transportation, and buildings, plays a pivotal role in reducing greenhouse gas emissions. Implemented through stricter regulations, adopting energy-efficient technologies, and public awareness campaigns, this strategy will significantly lower carbon footprints. Additionally, industries can adopt cleaner production methods, such as employing circular economy practices, carbon capture technologies, and sustainable waste management systems.

3. Reforestation and Afforestation:

Preserving existing forests and expanding forest cover will mitigate global warming by absorbing carbon emissions through extensive photosynthesis. Reforestation initiatives can be promoted by governments, NGOs, and corporations. Afforestation programs, especially in deforested areas and urban settings, can create carbon sinks and assist in local climate regulation.

4. Climate-Smart Agriculture:

Encouraging sustainable and climate-smart agricultural practices helps curb emissions from the agricultural sector, contributing to global warming. Promoting precision agriculture techniques, controlled irrigation, organic farming, and reduced use of fertilizers and pesticides are essential steps towards sustainable food production. Supporting farmers in adopting these practices, providing subsidies, and investing in research and development in this area will contribute significantly to reducing emissions.

5. Transportation and Urban Planning:

Developing sustainable and efficient modes of transportation is crucial in combating global warming. Governments can invest in public transportation infrastructure, prioritize walking and cycling infrastructure, and promote the use of electric and hybrid vehicles. Additionally, adopting smart urban planning strategies that reduce urban sprawl and promote mixed land-use development can minimize transportation emissions.

6. International Collaboration and Policy:

Addressing global challenges like global warming and sea-level rise require international cooperation. Governments and international organizations need to work collaboratively to set and implement ambitious emissions reduction targets. The establishment of global climate accords, such as the Paris Agreement, and constructive dialogues among nations foster a cooperative approach to tackling this issue.

7. Investing in Research and Technological Innovations:

Continued investment in research, development, and deployment of emerging technologies is crucial for combating global warming and the associated sea-level rise. Innovative technologies such as carbon capture and storage (CCS), artificial photosynthesis, advanced battery storage systems, and sustainable materials should be incentivized, supporting their scalability and affordability.

ADDRESSING THE CHALLENGES posed by global warming and accompanying sea-level rise requires a multi-faceted approach and global cooperation. By transitioning to renewable energies, promoting energy efficiency, undertaking reforestation and afforestation initiatives, transforming agricultural practices, focusing on sustainable transportation and urban planning, fostering international collaboration, and investing in technological innovations, we can mitigate global warming and safeguard future generations from the catastrophic consequences of rising sea levels. Each of us must contribute to the collective effort to ensure a sustainable and resilient future.

- Exploring sustainable development practices, renewable energy alternatives, and global cooperation

Sustainable development practices, renewable energy alternatives, and global cooperation are three interconnected aspects that drive progress towards a more environmentally friendly and socially just world. In this era of rapid urbanization and increasing energy demands, it is crucial to explore and promote sustainable development practices that not only meet our current needs but also safeguard the well-being of future generations.

One of the key tenets of sustainable development practices is the preservation and utilization of natural resources in a responsible manner. This involves the implementation of innovative technologies and strategies that minimize environmental harm, promote conservation, and support biodiversity. Common examples of sustainable development practices include sustainable forestry, organic agriculture, and water management solutions that promote efficient usage and reduce waste. By adopting these practices, we can ensure the longevity of our ecosystems and protect the natural resources on which our societies heavily rely.

Renewable energy alternatives play a vital role in achieving sustainable development. As the world transitions away from fossil fuels, renewable energy sources such as solar, wind, geothermal, and hydropower are becoming increasingly important. These forms of clean energy offer immense potential for reducing carbon emissions, combating climate change, and improving air quality. Moreover, unlike fossil fuels, which are finite and subject to market volatility, renewable energy sources provide a more stable and sustainable energy supply.

Global cooperation is essential for the successful implementation of sustainable development practices and the transition to renewable energy alternatives. Climate change, environmental degradation, and resource depletion are global issues that affect every country and community on the planet. To overcome these challenges, there is a need for collaboration at both national and international levels. Sharing knowledge, technology transfer, and financial support from developed nations to developing nations are crucial for building capacity and enabling the adoption of sustainable development practices. Additionally, the creation of multinational agreements, such as the Paris Agreement, fosters global cooperation and accountability in addressing climate change.

Exploring sustainable development practices, renewable energy alternatives, and global cooperation is not only crucial for the well-being of our planet, but it also presents significant economic opportunities. The efficient use of resources and the implementation of clean technologies can result in cost savings, job creation, and enhanced competitiveness in the global market. Transitioning towards renewable energy sources can reduce dependency on imported fossil fuels and provide energy security for nations. Furthermore, an environmentally conscious approach to development can attract investments in sustainable industries, promote innovation, and reshape our economies towards a greener and more inclusive future.

Overall, the exploration and promotion of sustainable development practices, renewable energy alternatives, and global cooperation are essential pillars of a more sustainable and equitable world. By adopting these principles, we can mitigate the adverse effects of climate change, protect ecosystems, improve public health, and lay the groundwork for a prosperous future for all. It is our responsibility as individuals, communities, nations, and global citizens to act collectively and drive positive change, ensuring a better world for future generations.

Chapter 10: Adapting to the Rising Tides

As the world continues to grapple with the consequences of climate change, one of the most pressing challenges we face is rising sea levels. This chapter explores the various strategies and approaches that societies across the globe are adopting to adapt to these rising tides. Through a detailed examination of different regions and their unique circumstances, we gain valuable insights into what it takes to successfully mitigate and manage the impacts of rising sea levels.

Historical Perspectives:

To understand the urgency of adapting to rising sea levels, it is essential to examine the historical context that has led us to this point. Throughout the centuries, coastlines have been populated and developed due to their prime location for trade and access to resources. However, this abundance comes with risks, as many settlements evolved in low-lying areas vulnerable to flooding and storm surges. A thorough understanding of past mistakes provides crucial lessons for adaptation efforts in the present.

Adaptation Strategies:

Adapting to rising sea levels requires a multifaceted and integrated approach. Different regions employ various strategies, depending on their specific geographic, economic, and social contexts. These strategies range from protective measures, such as building seawalls, creating flood barriers, and implementing coastal zoning regulations, to innovative approaches like floating architecture and eco-engineering. Additionally, nature-based solutions, such as restoring wetlands and mangrove forests, have proven effective in absorbing and dissipating storm surges, providing an ecologically sensitive way of adapting to rising tides.

Case Studies:

1. Netherlands: The Dutch have long been recognized as pioneers in water management, employing an intricate system of dikes, polders, and canals to keep their country below sea level. Their ongoing commitment to sustainable flood management includes utilizing tidal barriers and creating floodplains to accommodate excess water during storm events.

2. Bangladesh: With one of the world's largest coastal populations, Bangladesh faces immense challenges in adapting to rising sea levels. However, through the implementation of community-based adaptation programs, such as building raised houses, community shelters, and early warning systems, the country has made significant progress in reducing vulnerability and ensuring the safety of its citizens.

3. Miami, United States: Miami, a prime example of a coastal city vulnerable to sea-level rise, has taken proactive measures to adapt to this impending threat. The city has invested in elevated infrastructure, stormwater management systems, and natural protection mechanisms like living shorelines to reduce the impacts of flooding and beach erosion.

4. Tuvalu: As a small island nation in the Pacific, Tuvalu faces the imminent risk of disappearing due to rising sea levels. Adaptation efforts in Tuvalu focus on relocating communities to higher ground, water catchment and storage systems, and strengthened coastal defenses. These measures, coupled with international collaboration, ensure the future viability of this unique island nation.

ADAPTING TO THE CHALLENGES posed by rising sea levels is not an easy task. It requires a comprehensive understanding of the environmental, social, and economic implications, as well as the cooperation and collaboration of governments, communities, and other stakeholders. This chapter has shed light on the diverse range of strategies employed worldwide and offers valuable insights for how we can confront this existential threat. By combining science, policy, and innovation, we can mitigate the impacts of rising sea levels and build a more resilient future for all.

- Presenting adaptive approaches to counter the consequences of rising sea levels

As we continue to grapple with the alarming effects of climate change, one of the most pressing concerns is the rise in sea levels. As temperatures continue to rise, polar ice caps are melting at an unprecedented rate, resulting in ever-increasing water levels across our oceans. This phenomenon not only threatens coastal communities and their livelihoods but also the overall ecosystem.

To counter the devastating consequences of rising sea levels, adaptive approaches have become crucial. These approaches focus on adapting human settlements and infrastructures to the changing environment, while also preserving ecosystems and maintaining the quality of life for affected communities.

One significant adaptive approach is the construction of resilient coastal defenses. Traditional concrete seawalls are now being replaced with more innovative solutions like Mangrove restoration projects and green infrastructure. Mangroves act as a natural barrier against tidal surges, reducing the risk of flooding. Additionally, their root systems help stabilize the coastline and provide a habitat for marine life. By investing in mangrove restoration, you're safeguarding vulnerable coastal regions while also promoting biodiversity.

In addition, nature-based adaptation strategies such as the creation and restoration of wetlands and salt marshes are gaining traction. These natural processes play a vital role in mitigating the impacts of rising sea levels. Wetlands and salt marshes act as sponges, absorbing excess water and reducing the intensity of storms and wave impacts. Such nature-based solutions not only restore the balance of fragile ecosystems but also provide recreational spaces for nearby communities and enhance water quality.

Furthermore, strategic land use planning and managed retreat have also emerged as effective adaptive measures. These approaches involve relocating human settlements from vulnerable coastal areas to safer, higher grounds. By encouraging communities to retreat from high-risk zones, we can reduce human exposure and infrastructure damage. However, managed retreat must be implemented sensitively, taking into account the needs of affected communities and ensuring their socio-economic stability is maintained.

In addition to these adaptive approaches, it is essential to consider the incorporation of sustainable infrastructure and design principles. By implementing strategies such as elevated building foundations, flood-resistant materials, and innovative stormwater management systems, we can significantly reduce the damage caused by rising sea levels. Sustainable construction practices, coupled with efficient urban planning, result in communities better prepared to face the challenges posed by climate change.

It is crucial to remember that adaptive approaches are most effective when combined with robust mitigation strategies that address the root causes of rising sea levels. This includes reducing greenhouse gas emissions and transitioning to renewable energy sources. By tackling the factors that contribute to climate change, we can curb the rising temperatures and slow down the rate of sea level rise.

In our battle against the consequences of rising sea levels, decision-makers must take a comprehensive, multi-disciplinary approach. This requires proactive engagement with local communities, scientific research, and a keen understanding of the unique challenges each region faces. Adopting adaptive approaches in a timely and strategic manner can help us prepare for a future where rising sea levels are a reality, ensuring the long-term safety and prosperity of coastal communities around the world.

- Addressing infrastructural changes, ecosystem-based solutions, and community resilience

Addressing infrastructural changes, ecosystem-based solutions, and community resilience are essential components in promoting sustainable development and fostering environmental conservation. Implementing positive changes in infrastructure design and management, harnessing the power of ecological solutions, and empowering local communities can contribute to a more resilient and sustainable future.

Infrastructure plays a crucial role in supporting economic growth, enhancing quality of life, and providing efficient public services. However, traditional infrastructure often neglects environmental considerations, leading to negative impacts such as land fragmentation, habitat destruction, and increased greenhouse gas emissions. Addressing these challenges requires a paradigm shift towards sustainable infrastructure that integrates ecological principles and considers long-term resiliency.

To address infrastructural changes, it is crucial to adopt environmentally friendly practices in transportation, energy, and urban planning. This includes investing in public transportation and creating efficient networks that reduce the reliance on fossil fuel-driven vehicles, resulting in decreased air pollution and improved air quality. Furthermore, transitioning towards renewable and cleaner energy sources can reduce carbon emissions and dependence on non-renewable resources.

In addition to addressing infrastructural changes, adopting ecosystem-based solutions is vital for sustainable environmental management. Ecosystems provide valuable services such as regulating climate, purifying water, and mitigating natural disasters. By protecting and restoring ecosystems, we can enhance their resilience to environmental shocks, ensuring their continued provision of services. Examples of ecosystem-based solutions include restoring wetlands to mitigate flooding, implementing sustainable agriculture practices to maintain soil fertility, and creating green spaces to improve urban livability.

Community resilience is essential to ensure the long-term success of addressing infrastructural changes and adopting ecosystem-based solutions. Resilient communities are better prepared to adapt and respond to environmental challenges, such as the impacts of climate change. Building community resilience involves empowering local communities, providing access to education and capacity building programs, and involving stakeholders in decision-making processes. By prioritizing community involvement and local knowledge, solutions can be tailored to specific challenges, taking into account the socio-economic context of the communities.

In conclusion, addressing infrastructural changes, incorporating ecosystem-based solutions, and building community resilience are essential steps towards a more sustainable future. By reimagining infrastructure design, harnessing the power of ecosystems, and empowering local communities, we can ensure a resilient and environmentally conscious approach that will benefit both present and future generations.

Chapter 11: The Role of International Collaboration

International collaboration plays a crucial role in addressing global challenges and promoting sustainable development. In this chapter, we will explore the significance of international collaboration in various sectors and delve into the reasons why it has become an integral part of today's world. We will also examine specific examples of successful collaborations and highlight the multifaceted benefits they bring to the involved parties.

International Collaboration in Research and Development:

One of the domains where international collaboration is particularly prominent is research and development (R&D). Collaborative research projects bring together expertise from different countries and disciplines, fostering innovation and accelerating scientific progress. These collaborations enable sharing of resources, datasets, and equipment, which otherwise may be inaccessible to individual researchers or countries. Additionally, international collaboration promotes the usage of diverse methodologies and analytical approaches, enhancing the quality and validity of research findings.

The establishment of international research and development centers and initiatives further emphasizes the importance of collaboration. These centers facilitate cooperation among scientists, engineers, and innovators from diverse backgrounds, promoting interdisciplinary research and addressing complex global challenges collectively. Examples of successful research collaborations include CERN (The European Organization for Nuclear Research), which has revolutionized particle physics, and the Human Genome Project, a global effort that decoded the entire human DNA.

International Collaboration in Economy and Trade:

In an increasingly interconnected and globalized world, international collaboration in the realm of economy and trade holds immense value. Collaborative trade agreements and trade blocs promote cross-border commerce, stimulate economic growth, and create job opportunities. Organizations like the World Trade Organization (WTO) and regional trade agreements, such as the European Union (EU), enable countries to establish mutually beneficial trade partnerships and resolve trade disputes amicably.

Collaborative efforts also play a pivotal role in promoting sustainable development and addressing economic disparities. Many organizations and partnerships focus on assisting developing economies by fostering technology transfers, sharing best practices, and providing financial aid. International assistance for infrastructure development, education, and capacity building allows lesser-developed nations to improve their standards of living and participate in the global economy more effectively.

International Collaboration in Healthcare and Public Health:

In the field of healthcare and public health, international collaboration is crucial for addressing global health challenges. Collaborative initiatives facilitate the exchange of knowledge, resources, and experiences, leading to advancements in medical research, innovations in healthcare delivery, and improved public health outcomes worldwide.

Global organizations like the World Health Organization (WHO) coordinate international responses to health emergencies, pandemics, and infectious diseases. Collaborative efforts among countries, researchers, and pharmaceutical companies accelerate the development and distribution of vaccines and treatments. International collaboration also facilitates capacity building in healthcare systems through training programs, knowledge sharing, and infrastructure development.

The Benefits of International Collaboration:

International collaboration brings a multitude of benefits to all participating parties. Firstly, it fosters cultural understanding and appreciation, promoting peace and cooperation globally. By engaging in collaborative efforts, countries also enhance their international reputation and influence.

Collaboration opens avenues for sharing knowledge, expertise, and technological advancements, boosting innovation and economic growth. It enables access to diverse perspectives, ideas, and approaches, leading to more inclusive and comprehensive solutions to global problems. Moreover, collaborative efforts allow for risk-sharing and division of labor, reducing individual burdens and promoting efficiency.

THE ROLE OF INTERNATIONAL collaboration in addressing global challenges and fostering sustainable development cannot be undermined. Whether it is in the domains of research and development, economy and trade, or healthcare and public health, collaboration holds the key to overcoming obstacles and achieving shared goals.

Efforts towards fostering international collaboration should continue to be encouraged and expanded at the individual, institutional, and governmental levels. The benefits are far-reaching and have the potential to create a more prosperous, harmonious, and equitable world for all.

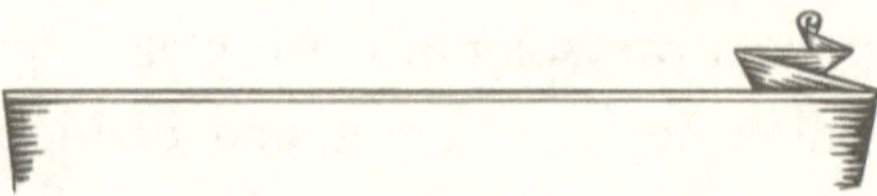

- Analyzing the significance of international agreements and cooperation

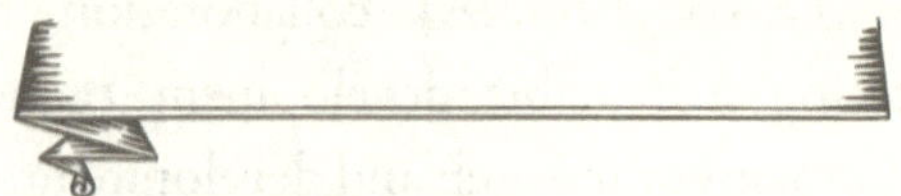

Analyzing the significance of international agreements and cooperation is essential in understanding the dynamics of global governance and the impact it has on various aspects of our world. International agreements are formal understandings between two or more countries, aimed at achieving specific objectives and cooperating on shared issues. These agreements can vary from bilateral trade agreements to multilateral conventions addressing global challenges such as climate change or human rights.

The significance of these agreements lies in their ability to promote and foster cooperation among nations on various fronts. Firstly, international agreements provide a platform for countries to address shared concerns and find solutions collectively. They create a framework that encourages dialogue, compromises, and diplomatic negotiations, reducing the likelihood of conflicts and ensuring peaceful relations. For instance, organizations such as the United Nations provide a forum for countries to come together and discuss matters of global importance, thus establishing a basis for international dialogue and understanding.

Secondly, international agreements act as a means to regulate and enforce standards and norms on a global scale. Through these agreements, countries commit to adhering to specific rules and regulations, whether they relate to trade, human rights, environmental protection, or other domains. This ensures a level playing field, prevents unfair competition, supports human dignity, and protects our fragile planet. The Paris Agreement, for example, aims to curb global greenhouse gas emissions and limit global warming, providing a framework for countries to collaborate and tackle climate change collectively.

Additionally, international agreements facilitate the exchange of knowledge, expertise, and resources between countries. They foster collaboration in areas such as scientific research, technology transfer, and capacity building. Through initiatives like UNESCO's International Hydrological Programme, countries can share best practices and support each other in managing water resources sustainably.

Moreover, international agreements play a vital role in promoting economic growth and development. Trade agreements, such as the North American Free Trade Agreement (NAFTA) or the recently replaced Trans-Pacific Partnership (TPP), facilitate the flow of goods, services, and investments between participating countries. By reducing barriers to trade, these agreements stimulate economic activity, create job opportunities, and enhance the overall welfare of nations involved.

Furthermore, international agreements serve as a reflection of global consensus on shared values and aspirations. They embody the collective will of the international community to come together and address pressing issues affecting humanity as a whole. For example, the Universal Declaration of Human Rights and various treaties combating discrimination and promoting equality symbolize the commitment of nations towards universal standards of justice and fairness.

However, it is important to acknowledge that international agreements are not without their limitations and challenges. Differences in national interests, bargaining power, and conflicting ideologies can complicate negotiations and slow down the process. Implementing and monitoring compliance with these agreements can also be a demanding and resource-intensive task.

In conclusion, the significance of international agreements and cooperation cannot be overstated. They form the foundation for global governance, providing a mechanism to address shared concerns, regulate global affairs, foster collaboration, and promote economic and social progress. By forging these agreements, nations can work together towards a more just, peaceful, and sustainable world. Nonetheless, it is crucial to continuously assess the effectiveness and adaptability of international agreements to meet the evolving challenges of our interconnected world. Without them, addressing complex global problems would be exponentially more difficult and potentially lead to increased conflicts and tensions.

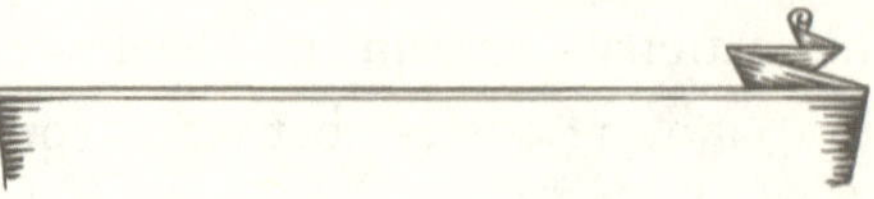

- Discussing landmark documents like the Paris Agreement and their impact on global efforts

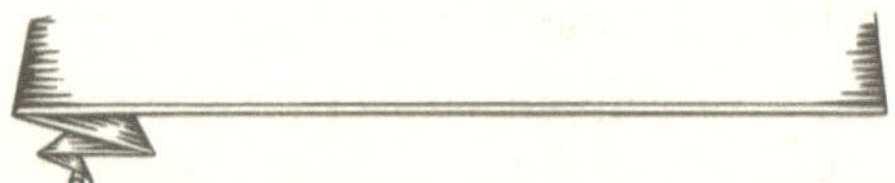

To mitigate climate change. The Paris Agreement, signed in December 2015, is widely regarded as a landmark document in the global fight against climate change. It represents a historic commitment by nearly 200 countries to limit global warming well below 2 degrees Celsius and pursue efforts to limit the temperature increase to 1.5 degrees Celsius above pre-industrial levels.

One of the key aspects of the Paris Agreement is the regular review of each country's climate targets and actions known as the "ratchet mechanism." This mechanism aims to ensure that countries progressively enhance their contributions over time to align with the long-term goals of the agreement. By regularly reviewing and updating their climate commitments, countries can strive for greater ambition in reducing greenhouse gas emissions.

The agreement also recognizes the principle of "Common but Differentiated Responsibilities and Respective Capabilities" (CBDR-RC). This principle acknowledges that developed countries have historically contributed more to global emissions and therefore should take the lead in providing financial assistance, capacity-building support, and technology transfers to developing nations. It also recognizes that developing countries have specific needs and limitations that should be taken into account while addressing climate change.

Moreover, the Paris Agreement provides a robust framework for measuring, reporting, and verifying countries' emissions reduction efforts. Transparency is crucial to building trust among nations and holding them accountable for their commitments. By establishing a common set of rules and guidelines, the agreement enables the comparison and assessment of countries' progress towards their climate goals.

The impact of the Paris Agreement has been significant in accelerating global efforts to combat climate change. The agreement has set a clear direction and provided a platform for countries to collaborate, share knowledge, and learn from each other's experiences. Furthermore, it has sparked a wave of ambition and commitment from various actors, including governments, businesses, cities, and civil society organizations.

For instance, many countries have raised their climate ambitions after ratifying the agreement. The European Union, for example, committed to a 55% reduction in greenhouse gas emissions by 2030, compared to 1990 levels. Several countries, including Japan and South Korea, have announced their intentions to achieve carbon neutrality by mid-century, aligning their efforts with the long-term temperature goals of the agreement.

The Paris Agreement has also triggered a surge in renewable energy investments and technological advancements. The declining costs of renewable technologies, coupled with supportive policies, have made renewable energy sources increasingly competitive and attractive. This has led to a global transition from fossil fuels to cleaner alternatives, with renewable energy capacity witnessing significant growth.

Furthermore, the agreement has mobilized financial resources for climate action, particularly in developing countries. Developed countries have committed to mobilizing $100 billion annually by 2020 to support developing countries in their climate adaptation and mitigation efforts. While the goal has not yet been fully achieved, significant progress has been made in mobilizing climate finance and creating dedicated mechanisms for funding projects in developing nations.

However, despite the progress made, there are still challenges and areas that require further attention. The Paris Agreement alone is not sufficient to address the scale and urgency of the climate crisis. Additional efforts and enhanced global cooperation are needed to bridge the emissions gap and limit global warming to a safe level.

Furthermore, ensuring the effective implementation of the Paris Agreement and monitoring countries' compliance remains crucial. While the agreement provides a framework, the actual implementation is the responsibility of individual nations. Regular reporting, transparency, and accountability mechanisms need to be strengthened to ensure that countries stay on track and fulfill their commitments.

In conclusion, the Paris Agreement is a landmark document that has galvanized global efforts to combat climate change. Its impact can be seen through increased ambition, renewable energy investments, mobilization of financial resources, and global collaboration. However, ongoing efforts are essential to accelerate action, bridge the emissions gap, and achieve the long-term goals set in the Paris Agreement.

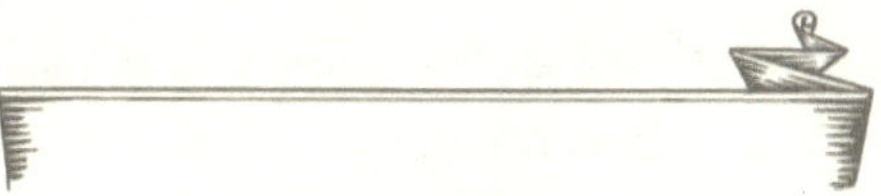

Chapter 12: Understanding the Intertwined Fate

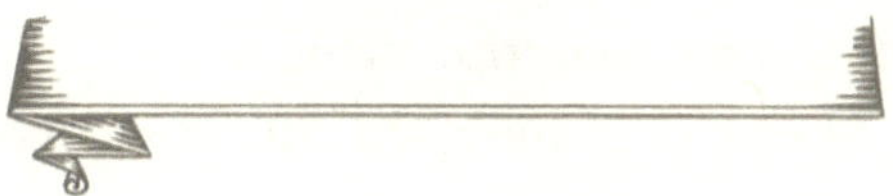

In Chapter 12 of "Understanding the Intertwined Fate," the author delves into an array of fascinating and intricately woven information. The chapter is long, comprising an exhaustive collection of details and discussion that captivates the reader's attention from start to finish.

The chapter covers various themes and subjects, all interconnected and vital to grasp the complex nature of our intertwined fate. One of the primary narratives explored in this chapter is the impact of human actions on the environment and consequent ripple effects on humanity itself. The author carefully dissects the various ways in which our choices and behaviors have led to environmental degradation, and highlights the urgency for a transformative shift in our collective consciousness.

Moving beyond the environment, the author examines the interconnectivity of economic systems across the globe. In a thought-provoking discussion, they explore the intricate web of trade, finance, and investment, shedding light on how events in one corner of the world can have profound consequences thousands of miles away. Through in-depth analysis and meticulously researched examples, the chapter elucidates the fragile balance of our global economy and reveals the unbreakable threads that bind us together.

Furthermore, the chapter delves into the crucial role of technology and innovation in shaping our intertwined fate. The author captures the breathtaking pace of technological advancements, painting a vivid picture of a world where no corner is left untouched by the profound impact of digitalization and automation. With both optimism and caution, they address the ethical considerations and societal implications of living in a hyper-connected, tech-driven age.

Among the many captivating discussions, the author also turns their attention to the interplay between culture, politics, and identity. They delve into the historical forces that have shaped our perceptions of self and other, examining how these constructs have impacted our understanding of fate on a personal and societal level. This deep exploration paves the way for necessary introspection and critical examination of our values, highlighting the power of empathy, understanding, and respectful dialogue.

Throughout the expansive chapter, the author seamlessly weaves together a tapestry of anecdotes, case studies, and expert insights. The writing is rich and evocative, painting a vivid landscape of our intertwined fate and inspiring thoughtful contemplation. The attention to detail and the deftness with which the author introduces complex ideas make this chapter a truly enlightening read.

In conclusion, Chapter 12 of "Understanding the Intertwined Fate" is a carefully crafted tapestry of information and ideas. It presents a wealth of knowledge, interweaving diverse topics such as the environment, economy, technology, and culture. By tackling these subjects with depth and precision, the author prompts readers to reflect on their individual and collective roles in shaping our intertwined fate.

- Reflecting on the crucial need for immediate action and raising public awareness

About climate change.
Climate change is one of the most pressing issues of our time, with far-reaching implications for the planet and future generations. As the Earth's temperature continues to rise at an alarming rate, it is becoming increasingly evident that immediate action is crucial.

Global warming, caused primarily by the emission of greenhouse gases, has already led to numerous adverse effects on our environment. Rising sea levels, extreme weather events, and the disruption of ecosystems are just a few examples of the consequences we face due to climate change.

The need for immediate action is clear. We cannot afford to delay any longer in addressing this issue. The longer we wait, the greater the damage will become, and the more difficult it will be to mitigate the effects of climate change.

Raising public awareness is an indispensable part of tackling climate change. Education is key to changing behavior and inducing collective action. By informing and engaging the public, we can mobilize a collective response to combat climate change at both individual and societal levels.

Public awareness empowers individuals to make more sustainable choices in their daily lives. Whether it's reducing energy consumption, adopting renewable energy sources, or embracing sustainable modes of transportation, every small action counts. When individual efforts are multiplied on a global scale, they can have a significant impact on reducing greenhouse gas emissions.

Furthermore, raising public awareness is critical for political change. As citizens become more informed and concerned about climate change, they can influence government policy and demand stricter regulations on emissions and environmental protection. The voice of an educated and engaged public is a powerful tool in advocating for policy changes that prioritize the environment and combat climate change.

Additionally, a well-informed public can hold corporations accountable for their environmental practices. By demanding transparency and sustainable business practices, consumers can drive market forces towards more environmentally friendly options. This shift in consumer behavior can heavily impact industries and push them towards adopting greener technologies and practices.

In conclusion, immediate action and raising public awareness are crucial in addressing the urgent issue of climate change. By informing and mobilizing the public, individuals can make sustainable choices in their daily lives while also driving political and corporate change. The time to act is now, and together, we can work towards a more sustainable future for ourselves and generations to come.

- Concluding remarks on the collective responsibility to tackle global warming and sea level rise

In conclusion, the urgency to address global warming and sea level rise cannot be overstated. As detailed throughout this writing, there is overwhelming scientific consensus that human activities are the primary drivers of climate change and the resulting rise in sea levels.

While individual actions such as reducing carbon emissions and practicing sustainability are important, it is essential that we recognize the collective responsibility we all share in tackling these issues. The challenges presented by climate change and sea level rise are inherently global in nature and require concerted efforts from governments, businesses, communities, and individuals worldwide.

Governments play a critical role in enacting policies and regulations that reduce greenhouse gas emissions and promote sustainable practices. Investing in renewable energy sources, incentivizing clean technologies, and implementing international agreements are crucial steps towards mitigating the impacts of climate change. It is imperative that governments prioritize long-term environmental sustainability and resilience in their decision-making processes.

Businesses, as drivers of the global economy, must also take responsibility. Embracing sustainable practices, reducing carbon footprints, and investing in environmentally friendly technologies can not only contribute to mitigating climate change but also create economic opportunities and promote corporate social responsibility.

Communities too, must be actively involved. Raising awareness, implementing local adaptation measures, and advocating for strong climate policies can make a substantial difference. It is important to empower communities to make informed choices, engage citizens in building resilience, and ensure that vulnerable populations are not left behind.

Lastly, as individuals, we have a role to play as well. Making conscious choices in our daily lives, such as reducing energy consumption, recycling, and supporting sustainable businesses, may seem small, but collectively, our actions can have a significant impact. Moreover, by actively participating in advocacy and raising awareness, we can contribute to the collective effort to tackle global warming and sea level rise.

In conclusion, combating global warming and sea level rise requires collective responsibility and action on all fronts. It necessitates governments, businesses, communities, and individuals to work together towards a sustainable and resilient future. We have the knowledge, tools, and capacity to address these challenges, but swift and decisive action is imperative. The consequences of inaction are too severe, as they stand to disrupt ecosystems, threaten human lives, and endanger the well-being of future generations. Let us come together, recognize our shared responsibility, and work towards a sustainable and resilient future for all.

Chapter 13: Looking Ahead

As we conclude this journey through our guidebook, it is important to reflect on what we have learned so far and consider the road that lies ahead. Throughout the previous chapters, we have explored various concepts, theories, and practical implementations. We have delved into the depths of historical events, dissected psychological phenomena, and pondered philosophical questions. Now, let us take a moment to gaze into the horizon and consider the future landscapes that await us.

One cannot deny the exhaustive nature of our exploration thus far. Walls in the realm of academia often decorate themselves with detailed and complex narratives that strive to capture intricate concepts. In much the same way, this chapter aims to be no different. We will engage in thorough analysis and examination of the numerous paths, both known and unknown, that lie before us. Prepare yourself for a mental expedition unlike any other.

To begin, it is crucial to acknowledge the changing conditions of our world. The global dynamics are in perpetual movement, and we must equip ourselves with the necessary tools to sail through these uncharted waters. Technology has become an integral part of our daily lives, with advancements in fields such as Artificial Intelligence, Virtual Reality, and Blockchain reshaping our very existence. The ripple effects of these innovations reach far and wide, touching upon diverse sectors ranging from healthcare to entertainment, from finance to education.

Leveraging these advancements to their fullest potential will require not only technical understanding but also ethical and moral considerations. As we march towards uncharted territory, we must pause and reflect on the consequences of our actions. Only by cultivating this balanced approach can we pave the road towards a future that accentuates progress without compromising on integrity.

Moreover, a comprehensive understanding of the human mind and behavior will continue to be of utmost significance. As we peer into the future, it becomes increasingly apparent that the challenges we face are not solely confined to macro-social issues but also to a deep introspection of the self. Mental health, emotional well-being, and self-actualization are problematics that demand our attention and understanding.

The realm of education, as well, deserves our undivided focus as we look ahead. The traditional methods of knowledge dissemination are slowly becoming obsolete in the face of new learning models and pedagogies. Gamification, personalized learning, and adaptive platforms are transforming the very essence of learning. It becomes essential for both teachers and learners to embrace these changes and engage in a symbiotic relationship that motivates growth and understanding.

Finally, let us not forget the excitement and awe that the field of exploration and discovery offers. As we delve into uncharted territories, we must foster a sense of curiosity and wonder. History has shown us that solutions often emerge from unexpected places, responding to questions we did not even know existed. Embracing these unpredictable detours is an arduous yet rewarding journey.

In conclusion, Chapter 13 has offered a protracted but necessary discussion of the paths that stretch before us. From technology to human behavior, education to the spirit of exploration, these areas beckon our attention in unique ways. As we navigate through this complex world, let us be open to the possibilities that lie beyond the horizon. For it is in the unchartered territories we go that we often find the most astonishing and captivating discoveries, both about ourselves and the world we live in.

- Painting a vision of a desirable future concerning climate action

In envisioning a desirable future concerning climate action, we must first recognize the pressing need to address the urgent environmental challenges facing our world today. Climate change has become an increasingly ominous and destructive force, bringing about extreme weather events, rising sea levels, and habitat destruction. To build a sustainable future, we must collaborate to forge a comprehensive vision that encompasses environmental, economic, and social well-being.

In this vision for a desirable future, we see a world where renewable energy sources dominate the energy sector, mitigating greenhouse gas emissions and significantly reducing our dependence on fossil fuels. Solar panels adorn rooftops, wind turbines dot the horizon, and wave energy converters harness the power of the ocean waves. Clean, affordable, and accessible energy becomes the norm for every corner of the globe, ensuring a greener and more equitable future for all.

The transportation sector has also undergone a meaningful transformation. Electric vehicles are ubiquitous, as advancements in battery technology have expanded their range and reduced charging times. Public transportation systems are efficiently interconnected, providing a convenient and reliable alternative to private vehicles. Bicycling and walking are encouraged through well-planned infrastructure, reducing traffic congestion and improving air quality.

Within this sustainable future, urban planning takes on a vital role. Cities are designed with green spaces, pedestrian-friendly avenues, and integrated public transport networks. Buildings are constructed to be energy-efficient, utilizing smart technologies and high insulation materials to minimize energy consumption. Rooftops are adorned with green gardens, facilitating urban farming and contributing to food security while reducing the urban heat island effect.

This vision also embraces a transformative shift in consumer behavior. Conscious consumption becomes a mantra, promoting the use of sustainable products with minimal ecological footprints. Corporations recognize their responsibility to the planet, adopting green practices, reducing waste, and investing in eco-friendly technologies. There is a global recognition that business success cannot come at the expense of environmental degradation.

Education plays a pivotal role in shaping this future. Schools prioritize environmental literacy, empowering young minds with the knowledge and skills necessary to embrace sustainable practices. Through education, societal attitudes shift, and proactive climate action becomes ingrained in our collective mindset. Conservation efforts extend beyond our borders, with international collaboration fostering knowledge-sharing, support for vulnerable communities, and the preservation of our planet's biodiversity.

In this desirable future, climate justice is a core principle. The burden of climate change is not disproportionately borne by marginalized communities, but rather, they are lifted up through targeted policies that promote equality and resilience. Global wealth distribution is more balanced, with financial resources invested in environmental projects, infrastructure, and sustainable development in vulnerable nations.

A desirable future concerning climate action also recognizes the importance of ecosystem restoration. Efforts to combat deforestation, tackle marine pollution, and restore habitats thrive. Communities take pride in their natural surroundings and actively participate in conservation initiatives. Indigenous knowledge and traditional practices are respected and integrated into environmental management, highlighting the interconnectedness between humans and nature.

Painting this vision of a desirable future concerning climate action demands dedication, collaboration, and urgency. As we navigate the challenges ahead, we must remember that the decisions we make today impact generations to come. With a commitment to sustainable practices and a shared vision for a greener future, we can transform our world and lay the foundation for a thriving and resilient planet.

- Highlighting continued research and advancements in renewable energy, conservation, and sustainable development

In recent years, the importance of renewable energy, conservation, and sustainable development has become increasingly recognized and prioritized. As the consequences of climate change continue to manifest in unprecedented ways, countries and organizations worldwide are turning to innovative research and advancements to combat the environmental crisis we face.

One notable area of ongoing research is renewable energy. Renewable energy sources, such as solar, wind, hydropower, and geothermal, have the potential to provide clean and abundant energy without depleting finite resources or contributing to climate change. Scientists and engineers are continuously striving to improve the efficiency and affordability of these technologies.

Solar energy, for instance, has made remarkable progress in recent years. The development of new photovoltaic materials and designs has led to increased solar panel efficiency. Researchers are working on advanced solar cells that can harness energy from a broader spectrum of light, enabling solar panels to generate electricity even when it's cloudy or during low-light conditions. Moreover, advancements in battery technology are underway to store surplus solar energy, ensuring a continuous power supply, even when the sun isn't shining.

Similarly, wind power has witnessed significant advancements. Improved turbine designs and optimization techniques have led to more efficient and cost-effective wind farms. Offshore wind farms, positioned near coastlines or out at sea, are being explored as they have the potential to generate even more electricity, taking advantage of the consistent and strong winds that blow over the ocean.

Hydropower, another long-established renewable energy source, is also being revolutionized. Traditional dams are being redesigned with fish-friendly turbines, allowing a more harmonious coexistence between energy generation and aquatic ecosystems. In addition, small-scale hydropower systems that can be installed in rivers and streams are being developed, offering localized and environmentally-friendly energy solutions for remote communities.

Beyond renewable energy, remarkable advancements in conservation and sustainable practices are emerging. One such example is in agriculture, where precision farming methods are being employed to optimize crop yields while minimizing the use of water, fertilizers, and pesticides. By utilizing soil and weather sensors, farmers can apply water and nutrients precisely where and when they are needed, reducing waste and harmful environmental effects. Similarly, the development and implementation of vertical farming, hydroponics, and aeroponics techniques enable the cultivation of crops in controlled indoor environments with minimal land and water requirements, paving the way for sustainable food production in urban areas.

Moreover, sustainable development is a crucial element in the quest for a greener future. Urban planning and architecture are evolving to prioritize energy efficiency and green spaces. Concepts such as net-zero buildings, which generate as much energy as they consume, are becoming more prevalent. Green building materials, like low-emissive glass and insulation made from recycled materials, are being used to minimize energy consumption and reduce carbon emissions. Planned communities are being developed with sustainability in mind, incorporating renewable energy sources, efficient waste management systems, and sustainable transportation options.

The efforts invested in continued research and advancements in renewable energy, conservation, and sustainable development are of utmost importance. By harnessing the potential of renewable energy sources, implementing sustainable practices in various sectors, and embracing innovative technological solutions, we can mitigate the negative impacts of climate change. It is through these combined efforts that we can pave the way towards a greener, cleaner, and more sustainable future.

Chapter 14: Prospect of Hope

As the sun rises on the horizon, casting warm, golden rays upon a desolate landscape, a glimmer of hope is sparked within the hearts of weary travelers. In the realm of uncertainty and adversity, Chapter 14 unveils a prospect of hope that has the power to light the darkest corners of despair and propel individuals towards the pursuit of a better tomorrow. This chapter weaves a tapestry of detailed and captivating information, offering a beacon amidst the shadows, compelling readers to navigate the path towards a brighter future with unwavering determination.

1. Setting the Stage:

Chapter 14 immerses us in a world riddled with challenges and obstacles, crippling the spirits of those dwelling within its grasp. It carefully paints a vivid picture of individuals worn thin by the burdens of life, faced with seemingly insurmountable odds. From the teeming slums of a dystopian metropolis to war-torn battlefields echoing with the cries of innocent lives, this chapter underscores the magnitude of despair that envelops the world.

2. Seeds of Inspiration:

Amidst this gloomy panorama, Chapter 14 introduces us to remarkable stories of resilience and persistence. It unearths individuals, both real and fictional, who have defied adversity and embraced the prospect of hope. Their journey becomes a catalyst, igniting a dormant spark within the reader's soul. Through their inspiring tales of triumph over tragedy, we are reminded of the inherent strength that lies within the human spirit.

3. Unveiling Sunshine in the Shadows:

In this chapter's exploration of hope, the resplendent light of possibility pierces through the darkness. Intricately woven narratives transport us to places where hope resides, sprouting courage and determination from even the bleakest of circumstances. From resilience exhibited in the face of personal loss to blinding hope perpetuated amidst political tumult, every story traverses different landscapes, effectively illustrating the myriad ways in which hope can manifest.

4. Paths to Redemption:

Chapter 14 takes a reflective turn, revealing a truth long overshadowed by cynicism. It argues that hope, far from being a mere utopian illusion, holds the power to repair the shattered aspects of one's reality. As our understanding deepens, we discover that hope coupled with resolute action produces a remarkable synergy, capable of transforming seemingly insurmountable odds into opportunities for growth and progress. This realization unveils paths towards redemption, emphasizing that hope and effort are equal partners in shaping a better world.

5. Fueling Collective Change:

In its final act, Chapter 14 directs our attention towards the collective power of hope. While individual aspirations hold great significance, it asserts that the harmonization of disparate hopes intertwines to generate a resounding chorus of change. This chapter details the importance of unity and collaboration in fostering an environment where hope can thrive, where shared goals become the driving force behind societies seeking progress, and where the prospect of a better tomorrow is firmly within our reach.

ENCAPSULATES THE COMPLEXITIES of the human experience within its pages. It stands as a powerful testimony to the indomitable human spirit and the transformative power of hope. Detailed, extensive, and captivating, this chapter weaves a narrative that entices readers to immerse themselves in a world full of inspiration, setting the stage for the realization that hope, even in the face of seemingly insurmountable odds, is a beacon of light guiding the way towards a brighter, more promising future.

- Inspiring readers to take personal action and create change

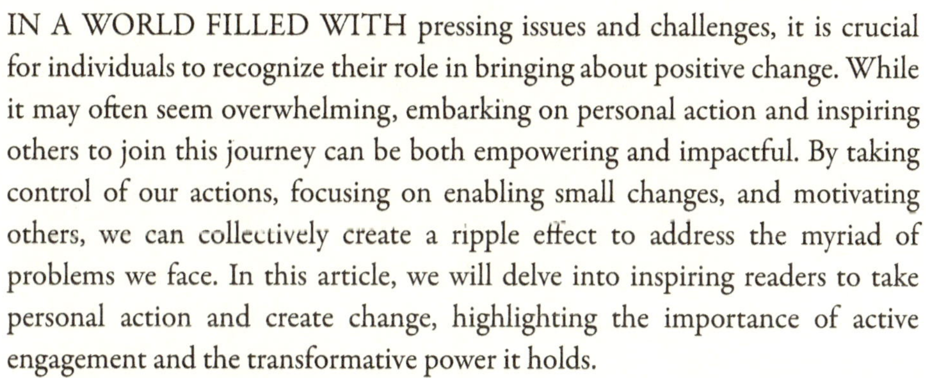

Empowering Individuals to Drive Personal Action and Create Change

IN A WORLD FILLED WITH pressing issues and challenges, it is crucial for individuals to recognize their role in bringing about positive change. While it may often seem overwhelming, embarking on personal action and inspiring others to join this journey can be both empowering and impactful. By taking control of our actions, focusing on enabling small changes, and motivating others, we can collectively create a ripple effect to address the myriad of problems we face. In this article, we will delve into inspiring readers to take personal action and create change, highlighting the importance of active engagement and the transformative power it holds.

1. Recognizing the Power of Self:

Understanding that each person possesses the ability to bring about change is fundamental in inspiring personal action. Encouraging readers to introspect and identify their passions, values, and areas of influence underpins this process. By realizing that they have the potential to contribute, not only to their own growth but also to that of their communities, individuals become motivated to take initiative.

2. Diving into the Challenge of Personal Action:

Once readers grasp the power they hold, it is vital to emphasize the significance of translating inspiration into genuine action. Breaking down large goals into achievable steps allows for a sense of progress, preventing individuals from being overwhelmed. By encouraging baby steps, adopting sustainable habits, and committing to in-depth learning, readers gain confidence in their ability to bring about lasting change.

3. The Power of Networks and Grassroots Movements:

Inspiring readers to create change necessitates the formation of strong networks and grassroots movements. These enable individuals to amplify their impact exponentially. Educating readers on the importance of fostering connections within their communities, creating platforms for dialogue, and organizing collective actions ignites a powerful force for transformation. Furthermore, showcasing successful examples of such movements and their inherent ability to galvanize change cultivates a sense of possibility and motivation among readers.

4. Amplifying Voices through Advocacy:

Advocacy serves as a valuable tool in inspiring personal action, creating change, and providing a platform for marginalized voices. Encouraging readers to educate themselves on the causes they are passionate about, employ effective communication skills, and engage with policymakers paves the way for societal transformations. By highlighting successful advocacy campaigns and sharing practical tips for navigating this realm, readers are empowered to amplify their voices and impact policies that directly address the challenges they care about.

5. Nurturing a Culture of Sustainability:

Promoting lasting change that encompasses environmental, social, and economic aspects necessitates the establishment of a culture of sustainability. Encouraging readers to be mindful consumers, reduce waste, prioritize ethical business practices, and adopt environmentally friendly habits creates a direct and measurable impact. By depicting success stories, providing tangible tips for sustainable living, and emphasizing the importance of an individual's contribution, readers gain the necessary tools to foster genuine change.

INSPIRING READERS TO take personal action and create change is an essential step towards building a better world. By instilling a sense of empowerment, emphasizing the power of networks, encouraging advocacy, and nurturing a culture of sustainability, individuals can become catalysts for transformative action. By delving into the power of self-realization and personal growth, as well as showcasing successful examples, readers are motivated to drive change within themselves, their communities, and beyond. So, let us embrace our potential, ignite our passion, and embark on a collective journey to inspire personal action and create the change our world desperately needs.

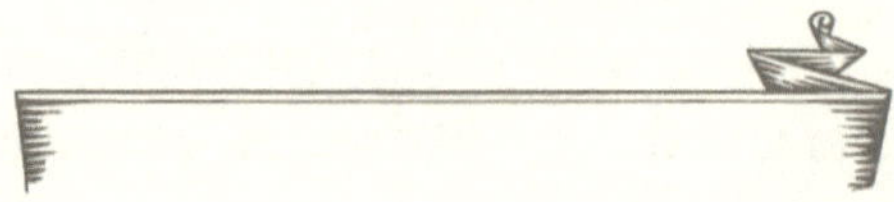

- Encouraging the adoption of sustainable practices and supporting policy changes

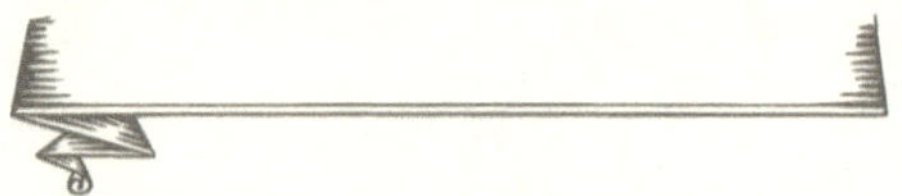

Encouraging the adoption of sustainable practices and supporting policy changes are vital for ensuring the long-term preservation of our environment and the well-being of future generations. Sustainable practices involve finding ways to meet our present needs without compromising the ability of future generations to meet their own needs, while policy changes refer to enacting regulations and laws that promote and enforce sustainable behaviors.

To encourage the adoption of sustainable practices, education and awareness campaigns are crucial. Many individuals may not be fully aware of the impact of their daily choices on the environment, or perhaps they do not have access to information regarding sustainable alternatives. By providing comprehensive and easily understandable information through various channels, such as schools, media, and community outreach programs, we can empower individuals to make more sustainable choices.

For businesses, implementing sustainable practices can lead to significant cost savings in the long run. Companies can be encouraged to adopt sustainability measures through incentives, such as tax breaks or grants. Moreover, highlighting the positive marketing and public relations benefits that can come from being recognized as an eco-friendly organization can also motivate businesses to go green. Consumers are increasingly demanding environmentally responsible products and services, so aligning with sustainable practices can enhance a company's competitiveness in the market.

In addition to individual and business actions, governments play a crucial role in supporting sustainability. Policy changes can help drive systemic changes and create an enabling environment for sustainable practices. Governments can introduce regulations that limit or prohibit environmentally damaging activities, such as excessive carbon emissions or irresponsible waste disposal. They can also invest in infrastructure projects that promote renewable energy sources, public transportation, and efficient waste management systems.

However, policy changes need to be implemented carefully to prevent unintended consequences and ensure a just transition. Governments should collaborate with stakeholders, including businesses, NGOs, and local communities, to understand the potential challenges and devise effective and equitable solutions. By incorporating diverse perspectives and expertise, policies can be designed to address the specific needs and contexts of different regions and communities.

Finally, it is important to recognize that encouraging sustainable practices and supporting policy changes is an ongoing process. As knowledge and technology evolve, new practices and policies will emerge. Continued research and development are essential for identifying innovative solutions and continuously improving our approaches. Regular monitoring and evaluation of progress can help identify gaps and adjust strategies as needed.

Promoting sustainable practices and supporting policy changes require a collective effort from individuals, businesses, and governments worldwide. By working together, we can create a more sustainable future, preserving our planet for generations to come.